SpringerBriefs in Space Life Sciences

AF803923

Series Editors

Günter Ruyters
Markus Braun
Space Administration
German Aerospace Center (DLR)
Bonn, Germany

The extraordinary conditions of space, especially microgravity, are utilized for research in various disciplines of space life sciences. This research that should unravel – above all – the role of gravity for the origin, evolution, and future of life as well as for the development and orientation of organisms up to humans, has only become possible with the advent of (human) spaceflight some 50 years ago. Today, the focus in space life sciences is 1) on the acquisition of knowledge that leads to answers to fundamental scientific questions in gravitational and astrobiology, human physiology and operational medicine as well as 2) on generating applications based upon the results of space experiments and new developments e.g. in non-invasive medical diagnostics for the benefit of humans on Earth. The idea behind this series is to reach not only space experts, but also and above all scientists from various biological, biotechnological and medical fields, who can make use of the results found in space for their own research. SpringerBriefs in Space Life Sciences addresses professors, students and undergraduates in biology, biotechnology and human physiology, medical doctors, and laymen interested in space research. The Series is initiated and supervised by Dr. Günter Ruyters and Dr. Markus Braun from the German Aerospace Center (DLR). Since the German Space Life Sciences Program celebrated its 40th anniversary in 2012, it seemed an appropriate time to start summarizing – with the help of scientific experts from the various areas - the achievements of the program from the point of view of the German Aerospace Center (DLR) especially in its role as German Space Administration that defines and implements the space activities on behalf of the German government.

More information about this series at http://www.springer.com/series/11849

Wolfgang Hanke • Florian P.M. Kohn •
Maren Neef • Rüdiger Hampp

Gravitational Biology II

Interaction of Gravity with Cellular Components and Cell Metabolism

Springer

Wolfgang Hanke
Institute of Physiology
University of Hohenheim
Stuttgart, Germany

Florian P.M. Kohn
Institute of Physiology
University of Hohenheim
Stuttgart, Germany

Maren Neef
Institute for Microbiology and Infection
Biology Tübingen (IMIT)
University of Tübingen
Tübingen, Germany

Rüdiger Hampp
Institute for Microbiology and Infection
Biology Tübingen (IMIT)
University of Tübingen
Tübingen, Germany

ISSN 2196-5560 ISSN 2196-5579 (electronic)
SpringerBriefs in Space Life Sciences
ISBN 978-3-030-00595-5 ISBN 978-3-030-00596-2 (eBook)
https://doi.org/10.1007/978-3-030-00596-2

Library of Congress Control Number: 2018960184

© Springer Nature Switzerland AG 2018
This work is subject to copyright. All rights are reserved by the Publisher, whether the whole or part of the material is concerned, specifically the rights of translation, reprinting, reuse of illustrations, recitation, broadcasting, reproduction on microfilms or in any other physical way, and transmission or information storage and retrieval, electronic adaptation, computer software, or by similar or dissimilar methodology now known or hereafter developed.
The use of general descriptive names, registered names, trademarks, service marks, etc. in this publication does not imply, even in the absence of a specific statement, that such names are exempt from the relevant protective laws and regulations and therefore free for general use.
The publisher, the authors and the editors are safe to assume that the advice and information in this book are believed to be true and accurate at the date of publication. Neither the publisher nor the authors or the editors give a warranty, express or implied, with respect to the material contained herein or for any errors or omissions that may have been made. The publisher remains neutral with regard to jurisdictional claims in published maps and institutional affiliations.

This Springer imprint is published by the registered company Springer Nature Switzerland AG
The registered company address is: Gewerbestrasse 11, 6330 Cham, Switzerland

Preface to the Series

The extraordinary conditions in space, especially microgravity, are utilized today not only for research in the physical and materials sciences—they especially provide a unique tool for research in various areas of the life sciences. The major goal of this research is to uncover the role of gravity with regard to the origin, evolution, and future of life, and to the development and orientation of organisms from single cells and protists up to humans. This research only became possible with the advent of manned spaceflight some 50 years ago. With the first experiment having been conducted onboard Apollo 16, the German Space Life Sciences Program celebrated its 40th anniversary in 2012—a fitting occasion for Springer and the DLR (German Aerospace Center) to take stock of the space life sciences achievements made so far.

The DLR is the Federal Republic of Germany's National Aeronautics and Space Research Center. Its extensive research and development activities in aeronautics, space, energy, transport, and security are integrated into national and international cooperative ventures. In addition to its own research, as Germany's space agency the DLR has been charged by the federal government with the task of planning and implementing the German space program. Within the current space program, approved by the German government in November 2010, the overall goal for the life sciences section is to gain scientific knowledge and to reveal new application potentials by means of research under space conditions, especially by utilizing the microgravity environment of the International Space Station (ISS).

With regard to the program's implementation, the DLR Space Administration provides the infrastructure and flight opportunities required, contracts the German space industry for the development of innovative research facilities, and provides the necessary research funding for the scientific teams at universities and other research institutes. While so-called small flight opportunities like the drop tower in Bremen, sounding rockets, and parabolic airplane flights are made available within the national program, research on the ISS is implemented in the framework of Germany's participation in the ESA Microgravity Program or through bilateral cooperations with other space agencies. Free flyers such as BION or FOTON satellites are used in cooperation with Russia. The recently started utilization of

Chinese spacecrafts like Shenzhou has further expanded Germany's spectrum of flight opportunities, and discussions about future cooperation on the planned Chinese Space Station are currently underway.

From the very beginning in the 1970s, Germany has been the driving force for human spaceflight as well as for related research in the life and physical sciences in Europe. It was Germany that initiated the development of Spacelab as the European contribution to the American Space Shuttle System, complemented by setting up a sound national program. And today Germany continues to be the major European contributor to the ESA programs for the ISS and its scientific utilization.

For our series, we have approached leading scientists first and foremost in Germany, but also—since science and research are international and cooperative endeavors—in other countries to provide us with their views and their summaries of the accomplishments in the various fields of space life sciences research. By presenting the current SpringerBriefs on muscle and bone physiology we start the series with an area that is currently attracting much attention—due in no small part to health problems such as muscle atrophy and osteoporosis in our modern aging society. Overall, it is interesting to note that the psycho-physiological changes that astronauts experience during their spaceflights closely resemble those of aging people on Earth but progress at a much faster rate. Circulatory and vestibular disorders set in immediately, muscles and bones degenerate within weeks or months, and even the immune system is impaired. Thus, the aging process as well as certain diseases can be studied at an accelerated pace, yielding valuable insights for the benefit of people on Earth as well. Luckily for the astronauts: these problems slowly disappear after their return to Earth, so that their recovery processes can also be investigated, yielding additional valuable information.

Booklets on nutrition and metabolism, on the immune system, on vestibular and neuroscience, on the cardiovascular and respiratory system, and on psycho-physiological human performance will follow. This separation of human physiology and space medicine into the various research areas follows a classical division. It will certainly become evident, however, that space medicine research pursues a highly integrative approach, offering an example that should also be followed in terrestrial research. The series will eventually be rounded out by booklets on gravitational and radiation biology.

We are convinced that this series, starting with its first booklet on muscle and bone physiology in space, will find interested readers and will contribute to the goal of convincing the general public that research in space, especially in the life sciences, has been and will continue to be of concrete benefit to people on Earth.

Bonn, Germany — Günter Ruyters
Bonn, Germany — Markus Braun
July 2014

DLR Space Administration in Bonn-Oberkassel (DLR)

The International Space Station (ISS); photo taken by an astronaut from the space shuttle Discovery, March 7, 2011 (NASA)

Extravehicular activity (EVA) of the German ESA astronaut Hans Schlegel working on the European Columbus lab of ISS, February 13, 2008 (NASA)

Preface

The first book on gravitational biology in our series "Springer Briefs in Space Life Sciences" entitled *Gravitational Biology I—Gravity Sensing and Graviorientation in Microorganisms and Plants* and published a few months ago had focused on gravitactic mechanisms of motile microorganisms and on gravitropic orientation of higher plants. Consequently, the book *Gravitational Biology II—Interaction of Gravity with Cellular Components and Cell Metabolism* digs one step deeper into the molecular and physiological basics by discussing the interaction of gravity with biological processes and subcellular structures.

In Chap. 1, the authors present evidence that during evolution obviously not only specific sensors for gravity sensing have developed, but that also cells and organisms without specific sensors for the direction of gravity are able to respond to an altered gravitational environment. Here, cell membranes—common to all cells—interact with gravity by changing its fluidity. Single molecules, isolated membranes, and cells have been studied to unravel the basic mechanisms of how gravity affects biological structures and processes. The consequences of these findings for life on Earth in general and specifically for human space travel are addressed. In a hierarchical model from single molecules, neuronal cells via the systems level, to the entire human brain, any impact of altered gravity conditions may lead to serious consequences up to changes in mental and cognitive capabilities of human beings.

Changes in membrane fluidity are also known to be relevant for pharmacology. Incorporation of hydrophobic and amphiphilic substances into membranes is clearly dependent on its fluidity. Since many pharmacologically relevant drugs belong to these substances, gravity may lead to changes in drug uptake into cells as could be shown for the model substance alamethicin. At the end of the chapter, the authors present a first functional model of a sensory system based on membrane thermodynamics.

Chapter 2 provides detailed insights into the influence of gravity and the consequences of the absence of gravity on gene and protein expression as well as cell metabolism. Again, not the directional information of the gravity vector sensed by organisms via very specific mechanisms and realized in a dedicated and complex signal transduction chain like in gravitaxis and gravitropism is in focus, but the more general effects on molecular and metabolic processes of cells are described. Mainly based on space experiment data from callus cultures of *Arabidopsis thaliana*, the present knowledge in the field of gravity-affected cell metabolism is summarized. Especially gravity-induced changes in the flow of calcium ions and of reactive oxygen species (ROS) such as hydrogen peroxide and thus their role as second messengers in metabolic pathways and the importance for cell signaling are discussed. The authors conclude by providing a model for the early signaling events that are initiated by altered gravity stimulation combining the changes in Ca^{2+} and hydrogen peroxide with fast responses in gene expression, protein modulation, and metabolic pathways.

All in all, this book nicely complements the previous publication *Gravitational Biology I—Gravity Sensing and Graviorientation in Microorganisms and Plants* by providing a view on gravity-induced effects on cellular structures and biological processes—from molecules, cell membranes, and second messengers, via gene and protein expression, to cellular functions.

Bonn, Germany
August 2018

Günter Ruyters
Markus Braun

Acknowledgements

We gratefully acknowledge the financial support by Deutsches Zentrum für Luft- und Raumfahrt (DLR). We are also deeply indebted to PD Dr. Markus Braun for many helpful suggestions in general, as well as to Ulrike Friedrich for her help in parabolic flight campaigns, both from DLR Space Administration. Margret Ecke (University of Tübingen) was a continuously and highly motivated member of the research group from the very beginning. Also in general, we appreciate the support by so many members of the German space industry over the years (ERNO, EADS Astrium, Airbus).

Contents

Chapter 1
Interaction of Gravity with Cellular Compounds

Abstract During evolution, the majority of organisms have developed specific sensors for gravity, the only constant environmental cue on earth. Nevertheless, a variety of gravity effects on molecular, cellular, and physiological level has also been reported in single-cell organisms and cell types of plants and animals which do not seem to possess specific sensors. We have found that the cellular membrane, common to all cells, itself is interacting with gravity by changing its fluidity. Thus, it delivers a basic mechanism for gravity perception for all existing cells and living systems. In the following, we discuss the physical principles and the consequences of our findings for membrane-bound processes, for life on earth, and for manned space travel. In addition, a first model is proposed, how a sensor system for gravity based on membrane thermodynamics could be structured.

Keywords Neuronal cells · Gravity perception · Membrane fluidity

1.1 Introduction

Life on earth has evolved under constant 1g gravity conditions, and most organisms have developed specific organs or mechanisms to sense the direction of gravity, to be used, i.e., for orientation in the environment. Such mechanisms exist down even to the level of single cells. However, more general nondirectional gravity effects on molecular, cellular, and physiological processes have been found in many cell types, which are not known to be equipped with specific gravity-sensing mechanisms. This finding gives rise to the question of more basic mechanisms of gravity sensing on the molecular, membrane, or cellular level (Häder et al. 2005).

Especially neuronal cells of higher vertebrates are not known to use specific organelles for gravity sensing; however, in case they would be able to sense gravity with other mechanisms, this would possibly have significance for man living in space. To address these questions, experiments have to be performed, studying possible functions of single molecules and cellular membranes in gravity-sensing processes, as there are, for example, the generation and propagation of action potentials and the

© Springer Nature Switzerland AG 2018
W. Hanke et al., *Gravitational Biology II*, SpringerBriefs in Space Life Sciences,
https://doi.org/10.1007/978-3-030-00596-2_1

reaction of chemical synapses in response to changing gravity conditions. In a hierarchical model of neuronal systems, from single molecules to the entire human brain, any gravity dependence in the lower levels would unavoidably lead to serious consequences up to changes in mental and cognitive capabilities of humans in space.

Cells including neuronal cells, membranes, and even molecules are highly complex structures and thus have to be looked at within the framework of nonlinear thermodynamics. They can be addressed as either internally or externally energized excitable media giving rise to the consequence that they are critically depending on small external forces to which gravity belongs by definition. This again supports the necessity of experiments with neuronal structures under conditions of changing gravity conditions.

Additionally, studying the complex neuronal systems, experiments with simpler model systems, for example, with plain lipid vesicles or oscillating chemical reactions, and computer-based simulations, can be used to explore the gravity dependence of basic biochemical and biophysical processes.

Whereas vacuum and radiation conditions as given in space can be simulated on earth within some limits in sufficient quality and duration, platforms to produce microgravity are either of short duration (seconds to minutes) or extremely expensive. The best example is the ISS, by sure the most famous microgravity platform available today. In principle, experiments can be done here at timescales up to years; however, the preparation of the experiments is extremely time-consuming, highly complex, and expensive and the time needed to organize such an experiment can be some years, and the safety regulations are strict.

Fortunately, molecular, membrane, and neuronal cellular reactions to gravity changes have been found to be at least partially fast, usually in the millisecond or second to minute range. Thus, mainly short-duration platforms like the drop tower, parabolic flights, and sounding rockets have been conducted in the field of gravitational biology research with membranes, single cells, and action potentials.

All microgravity research must be accompanied by proper ground controls and hypergravity experiments. Producing hypergravity is possible with centrifuges even at long timescales and with relatively big experimental setups. If necessary and available, data from ground controls and hypergravity experiments will be presented together with results from microgravity experiments in this chapter.

1.2 Structure and Function of the Neuronal System

Although this chapter mainly focuses on the gravity dependence of molecules, membranes, and cells, a short walk through the systematic structure of central nervous tissue and its connections will help to understand the possible consequences of the gravity dependence in those.

The CNS tissue is a highly complex structure, which for our purposes can be seen as a hierarchical system, reaching from single molecules and cells to the entire brain, as depicted in a simplified model in Fig. 1.1. We start from the presence of molecules

Fig. 1.1 Hierarchical structure of the human CNS

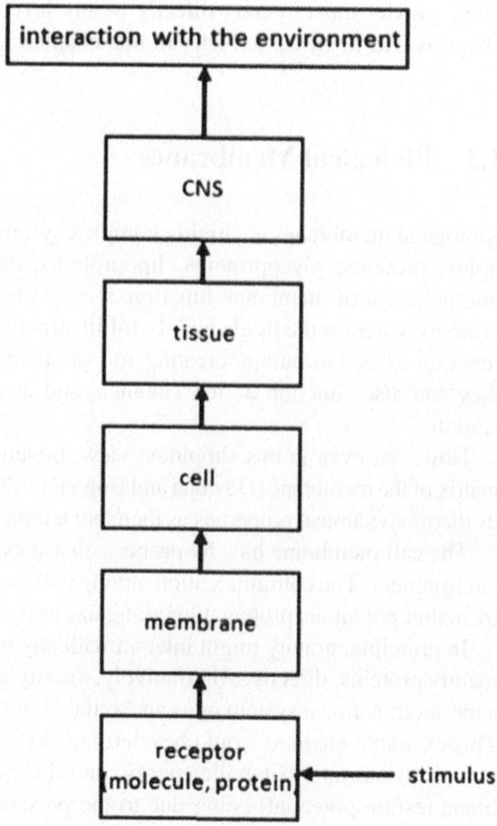

such as lipids and proteins, which together form the plasma membrane and the membranes of organelles. Membranes create closed biochemical systems, cells, or organelles, which are connected to build tissue and higher structures, as there is finally the entire brain. These cells interact with the environment and among each other by electrical and chemical communication. Action potentials, which are the central electrical communication units in neuronal systems are produced by neurons and propagate along axons to other cells. At the connection between an axon and the next cell, signals are transferred via chemical synapses.

Gravity might interact with such a system on all levels of complexity including specific gravity-sensing structures, although it is widely assumed that the human brain itself has no specific gravity-sensing structures.

The main sensory input for gravity, position in space, and acceleration conditions to the human brain is the vestibular organ of the inner ear and internal receptors, which are not part of the CNS, and the visual system. Especially the inner ear and the eyes deliver just input channels to the brain. All parts of the human CNS itself are not known to be equipped with any specific gravity-sensing structures. Nevertheless, part of them including single neurons have been shown to react to gravity changes.

Thus, gravity must interact directly at any level with the system via other mechanisms as will be discussed later in this chapter.

1.3 Biological Membranes

Biological membranes are highly complex systems themselves, mainly composed of lipids, proteins, glycoproteins, lipoproteins, and other molecules. The classical interpretation of membrane function assigns the functional properties mainly to the proteins, whereas the lipids mainly fulfill structural tasks. Proteins can, for example, function as active pumps creating ion gradients by consuming metabolic energy; they can also function as ion channels and as receptors for chemical or physical stimuli.

However, even in this simplified view, proteins are always embedded in the lipid matrix of the membrane (Dowhan and Bogdanov 2002). Thus, they will be depending on its thermodynamical properties as there are temperature, fluidity, and other parameters.

The cell membrane has, by proper constructs, to interact with other cells and the environment. The communication among cells in neuronal tissue is mainly organized by action potentials propagating along axons and chemical and electrical synapses.

In principle, gravity might interact with any of the components of the membrane, mainly proteins, directly. Alternatively, gravity might interact with the membrane as a thermodynamical system or as an excitable medium and thus change its properties. This excitable medium would be energized by ion gradients created by the action of pumps consuming metabolic energy, and the electrical potential (mainly the membrane resting potential) being due to the presence of ion gradients and of selective ion channels, which can be calculated by the Goldman equation (Goldman 1943). In such an interpretation, protein function might be changed by physical membrane properties, especially the membrane fluidity in a secondary step (Aloia and Boggs 1985).

Indeed, it has been shown in numerous experiments that the physical state of the membrane is an important factor to modulate protein function. For example, the kinetics of the pore-forming properties of the nicotinic acetylcholine receptor is depending on the physical state of its environment (Zanello et al. 1996).

1.4 Neuronal Cells

Types of cells in the CNS are mainly neurons and glial cells. Neurons are the main signal processing units of the brain, and they are interconnected via dendrites, axons, and chemical synapses. Additionally, electrical synapses are directly connecting cells in the CNS.

Action potentials (APs) are generated within a neuron by integration over all input signals and then, via its axon, are delivered to other cells in the CNS. At the interface between axon and the next cell, chemical synapses are the mechanistic units organizing signal transfer.

Again, gravity might interact with different structures of such a neuron, even including the intracellular and extracellular matrix. Axons additionally can be seen as cable-like structures in which action potentials propagate. The geometrical and electrical properties of such axons, which are influencing the properties of propagating APs, also might be gravity sensitive. Using the terminology of excitable media, APs are furthermore propagating waves in an excitable medium, which are gravity dependent by definition via the interaction of such media with small external forces.

1.5 Thermodynamics of Neuronal Systems

In short, biological systems including all structures discussed up to now might be seen as excitable media (Epstein and Pojman 1998; Sagués and Epstein 2003; Tabony 2006). Such a medium is based on the following physical needs:

- It must be thermodynamically open.
- Mass and/or energy transfer through the system should be present.
- The system should be far away from equilibrium.
- Feedback must be present within it.
- The system must be composed of a large number of small units.

Such a system can be energized internally or externally. In neuronal structures, metabolic energy usually will be the main source. As a consequence of the above mechanism, the system has emerging properties as there are among others:

- Self-organization
- Pattern formation
- Oscillations
- Propagating waves

and the system in its properties is critically dependent on small external forces including gravity as a small physical force. Of course, an excitable medium has to obey the first and second law of thermodynamics. Usually excitable media are described in terms of nonlinear thermodynamics. The mathematical presentation is given by coupled sets of nonlinear differential equations or possibly for computer-aided simulation by cellular automata (Acedo 2009; Wolfram 2002) or other formalistic systems. Applying the just stated theoretical approaches to neuronal systems, it becomes obvious that the brain, parts of it, but also single cells can be interpreted as excitable media (Wiedemann and Hanke 2002); however, plain membranes or vesicles with identical internal and extracellular aqueous solution are not excitable media, as they are energetically close to equilibrium systems with only thermodynamical fluctuations.

Another important fact about neuronal systems (biological systems in general) is that the biological membranes of neurons may be described as two-dimensional

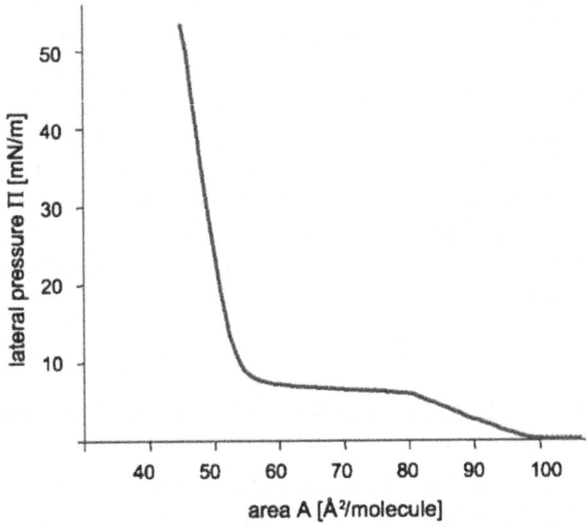

Fig. 1.2 Monolayer on a film balance. The film pressure is plotted as function of the area per lipid molecule. The plateau is due to the main phase transition of the lipid

systems, which are defined by temperature, lateral pressure, and area. Accordingly, at constant temperature as usually given in neuronal systems, such a membrane can be described by a graph plotting lateral pressure against area. To do so for a real bilayer mechanistically is very difficult, but using a lipid monolayer on a film balance as a model system, such curves can be measured easily as shown in a simplified example in Fig. 1.2. Here, a graph showing the film pressure as a function of area is given for a plain lipid with a defined phase transition temperature, as can be seen in the graph.

According to the Singer-Nicolson fluid mosaic model (Singer and Nicholson 1972), biological membranes are usually in the fluid state but might be shifted along the curve describing its state. As biological membranes are of highly complex structure, composed of different lipids, proteins, and other molecules, the situation usually will not be as simple as given in Fig. 1.3. However, even in real membranes, phase transitions might occur, and any membrane and more specific protein function will depend on the actual position of the system membrane in the physical space.

1.6 Interaction of Gravity with Single Molecules

A very basic question in the field of gravity research is whether gravity directly changes the properties of single molecules with relevant biological function. One of the main functional units of neuronal membranes is ion channels formed by different proteins. They are involved among others in ion homeostasis, signal transduction, and the generation of action potentials. Typical structural elements of such ion channels are alpha-helical and beta-sheet parts of the molecule spanning the membrane. Typical properties of an ion channel are the opening and closing times and the

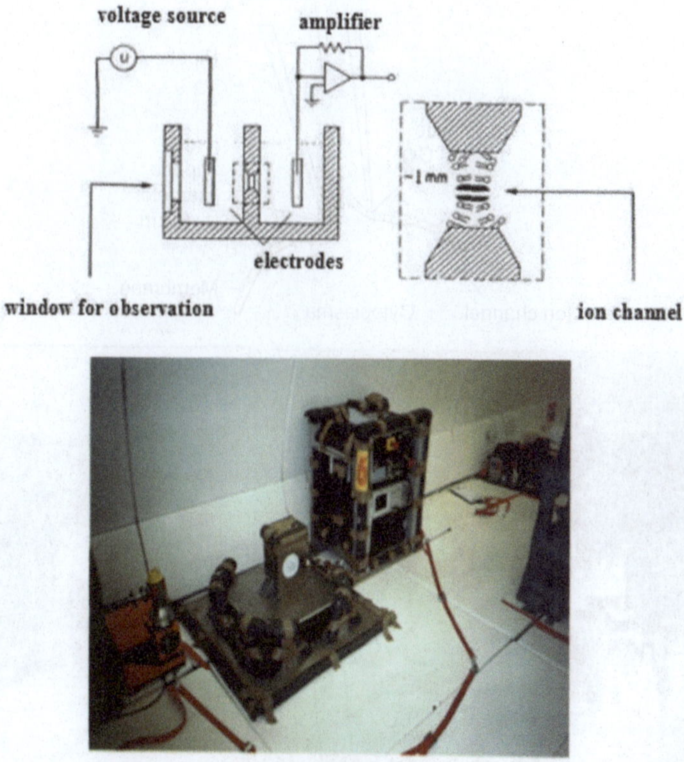

Fig. 1.3 The sketch in the upper part shows the principals of a bilayer experiment; below, a photo of a setup for bilayer experiments used in parabolic flight missions is shown [modified from Wiedemann et al. (2011)]

open-state probability and its selectivity. Additionally, it must be considered by which parameters the ion channel is controlled, usually by ligands or by membrane potential.

Technically, the abovementioned parameters of ion channels can be investigated best by electrophysiological techniques like the patch-clamp technique (Hamill et al. 1981) or the reconstitution of proteins into planar lipid bilayers (Hanke and Schlue 1993). With such techniques either single-channel studies or voltage- or current-clamp experiments can be done.

In Fig. 1.3, the basics of a bilayer experiment are shown together with a photo of a setup used in parabolic flight experiments. In Fig. 1.4, the patch-clamp technology is depicted, and the technology used for patch-clamp experiments under microgravity in a parabolic flight mission is shown (Meissner and Hanke 2002). Experiments utilizing both techniques were done in parabolic flight campaigns and in the drop tower. According to the mechanical sensitivity of patch-clamp experiments and the need for sensitive handling by the scientist in parabolic flight campaigns, only a few

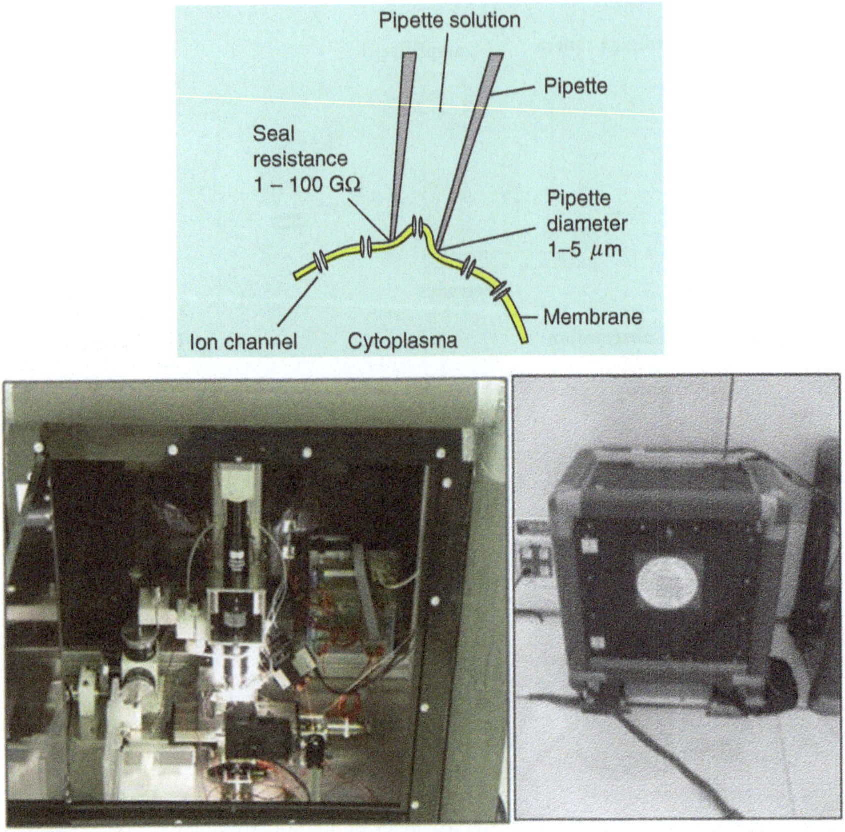

Fig. 1.4 The upper part shows the basics of the patch-clamp technology. The photo in the lower part shows an older setup as used in parabolic flight missions. This setup also has been adapted to the drop tower [modified from Wiedemann et al. (2011)]

such experiments were performed, and data are still rare. With the development of the planar patch-clamp technology, semiautomatic equipment for patch-clamp experiments became available.

Due to the much easier handling and to advances in mechanical stability, later, mainly this technique was used in microgravity experiments in parabolic flight missions (i.e., Wiedemann et al. 2011). The basics of this technology are depicted in Fig. 1.5, and a photo of a real setup is shown in Fig. 1.6. Here a commercial setup, the Port-a-Patch system from Nanion®, was adapted for the experiments.

Regardless of the advances in technology, single native ion-channel electrophysiological experiments have been found to be extremely difficult to conduct under microgravity conditions (Kohn 2010); thus, published results are rare. Some experiments with the model pore alamethicin and with porins, however, have delivered useful data in the field. The alamethicin pore is created by bundles of alpha helices in a barrel-like fashion (Boheim 1974; Hall et al. 1984; Woolley and Wallace 1992;

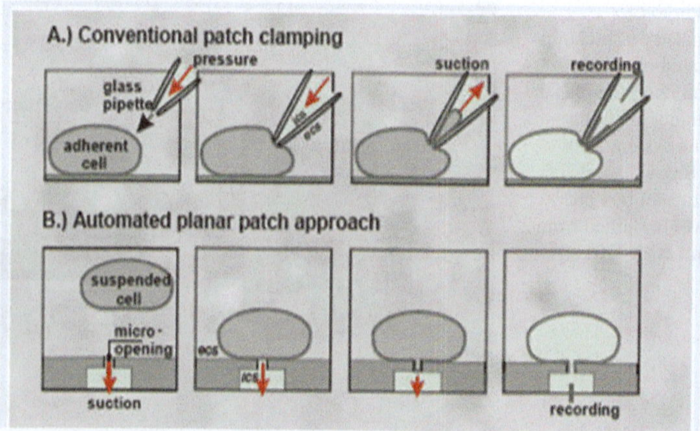

Fig. 1.5 This sketch compares the classical patch-clamp technique to the planar patch-clamp technology tower [modified from Wiedemann et al. (2011)]

Leitgeb et al. 2007) and is a well-suited and deeply studied model system for ion channel research (Boheim et al. 1983, 1984; Cafiso 1994).

In direct single-channel studies in planar lipid bilayers, it was found that the activity of alamethicin-induced pores is gravity dependent (Hanke 1995; Klinke et al. 2000; Wiedemann et al. 2003). The overall pore activity increases via increasing gravity. The same was found for a pore-forming porin reconstituted into a planar lipid bilayer (Wiedemann et al. 2003, 2011). An example is shown in Fig. 1.7. Single alamethicin data are shown in Fig. 1.8.

It should be mentioned here, however, that optical techniques are increasingly been used in the field and might be the future of investigating functional properties of ion channels. Such new optical techniques possibly in some time can be used in the field of "live sciences under space conditions" and will help to obtain the required data.

In addition, in experiments using oocytes with overexpressed epithelial sodium channels, evidence was found that these channels close toward reduced microgravity and open upon hypergravity (Richard et al. 2012). The experiments were performed in a parabolic flight mission. Measurements of whole-cell currents utilizing the patch-clamp technique resulted in a more complex dependence of membrane permeability on gravity (Fig.1.9). This is most probably given by the fact that whole-cell currents usually are a complex mixture originating from a variety of active channels and transporters. In laboratory experiments, usually pharmacological interventions are done, for example, to block part of the membrane conductance of the cell by proper drugs (Wiedemann et al. 2011).

The use of most of the relevant drugs to block certain ion-channel populations or transporters (i.e., to block sodium channels by TTX), unfortunately, is not allowed in parabolic flights, as usually they are significantly toxic. This violates safety regulations given for parabolic flights (Novespace 2009) and air travel in general.

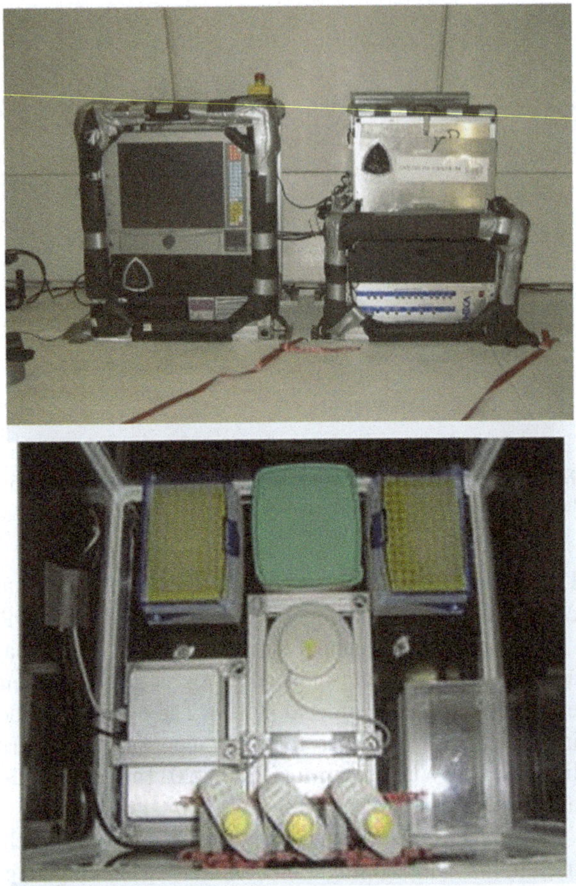

Fig. 1.6 Photo of a semiautomated patch-clamp setup for parabolic flight missions utilizing the planar patch technique with the Port-a-Patch system from Nanion®. Below, details of the experimental box are shown tower [modified from Wiedemann et al. (2011)]

Another approach to measure currents across cell membranes under variable gravity conditions was chosen by Richard et al. (2012). They recorded currents across oocyte membranes in a parabolic flight experiment with a specifically designed setup, and they found a significant reduction of current under microgravity. This could be interpreted as closing of ion channels in the membranes of the oocytes and thus would be consistent with our results.

Additionally, from a variety of experiments in so-called simulated microgravity, using clinostats or random positioning machines, but also mechanical unloading, data has been published suggesting a reduced ion-channel activity under microgravity. According to our statement about the physical questions behind these technologies, we will not comment the results in detail here but just point out that, when being used, they support the above given results from real microgravity experiments.

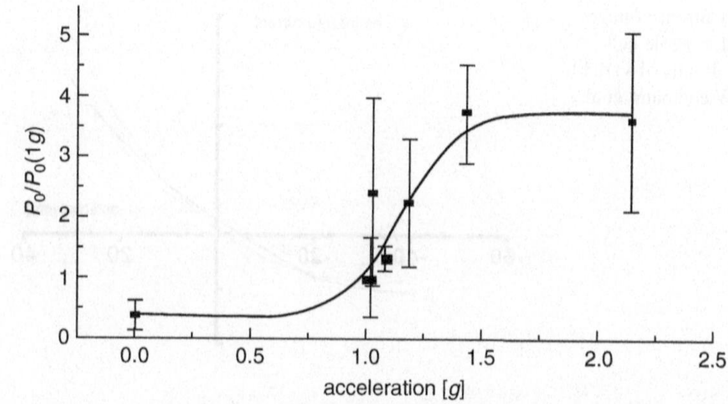

Fig. 1.7 Dependence of single porin pore fluctuations on gravity. Traces of alamethicin fluctuations are shown at different gravity levels [modified from Goldermann and Hanke (2001)]

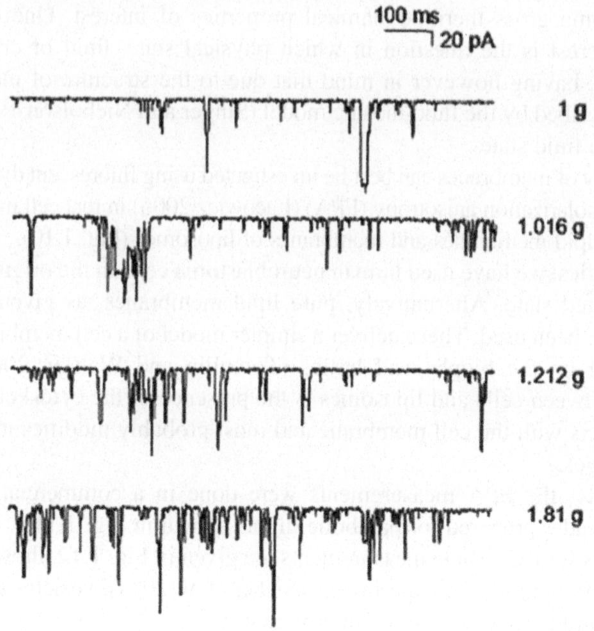

Fig. 1.8 The open-state probability of alamethicin pore fluctuations on gravity is shown. Obviously, pore activity increases toward higher gravity levels (Wiedemann et al. 2011)

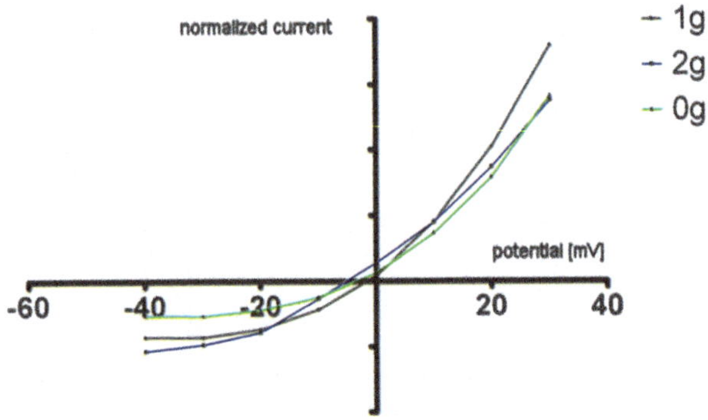

Fig. 1.9 Current voltage relation of a whole cell under conditions of variable gravity (Wiedemann et al. 2011)

1.7 Interaction of Gravity with Membranes

Biological membranes, as stated already, are complex constructs mainly from proteins and lipids. Aside from the functional properties as implemented by the proteins, they have some gross thermodynamical properties of interest. One of the major points of interest is the question in which physical state, fluid or crystalline, the membrane is, having however in mind that due to the structure of membranes, as basically described by the fluid mosaic model (Singer and Nicholson 1972), they are usually in the fluid state.

The fluidity of membranes can best be investigated using fluorescent dyes to measure fluorescence polarization anisotropy (FPA) (Lacowicz 2006) in real cell membranes but also in plain lipid membranes and membranes of liposomes (Fig. 1.10).

In our studies, we have used human neuroblastoma cells in the original and in the re-differentiated state. Alternatively, pure lipid membranes, as given by artificial vesicles, have been used. These deliver a simpler model of a cell membrane allowing concentrating on the biophysical basics (Torchilin and Weissig 2003). Another difference between cells and liposomes is the presence of the cytoskeleton in cells, which interacts with the cell membrane and most probably modifies its response to gravity changes.

Technically, the FPA measurements were done in a commercial multimodal, multi-plate reader prepared for parabolic flights shown in Fig. 1.11.

The results for the vesicle measurements are given in Fig. 1.12, those for the cells in Fig. 1.13. In both sets of experiments (Sieber et al. 2014), vesicles and real cells, membrane fluidity increases under microgravity.

An exception is given, when liposomes at a temperature in the range of the phase transition temperature of the lipid are used. Here, no fluidity changes due to gravity changes were found. This is due to the fact that membranes are highly compressible in this state.

In an additional set of experiments, the size of plain lipid vesicles was investigated in the drop tower using the light-scattering technology. In these experiments, it was found that the vesicles became slightly bigger under microgravity

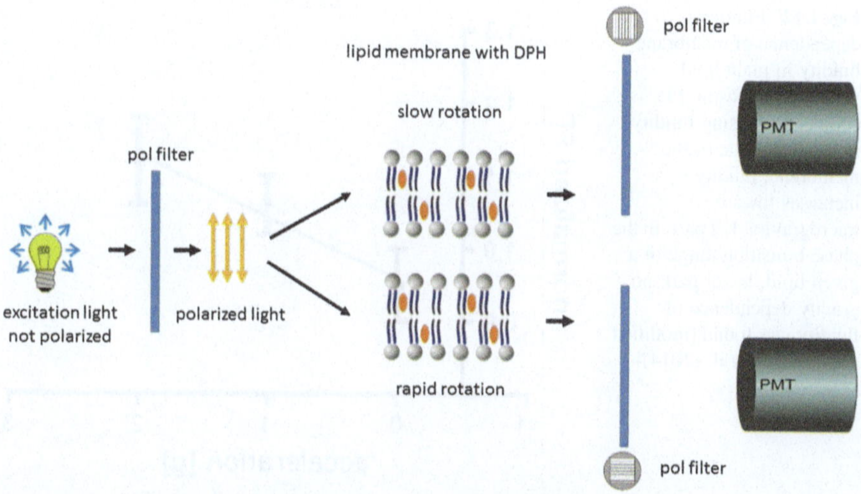

Fig. 1.10 Technology of the fluorescence polarization anisotropy measurements using DPH as fluorescent dye [modified according to Zhao and Lappalainen (2012)]

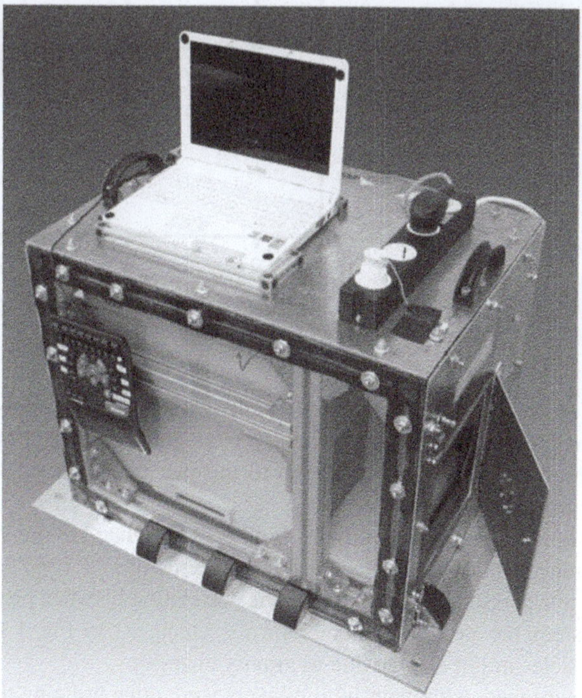

Fig. 1.11 A setup build around a 96-well universal plate reader for parabolic flight missions is shown

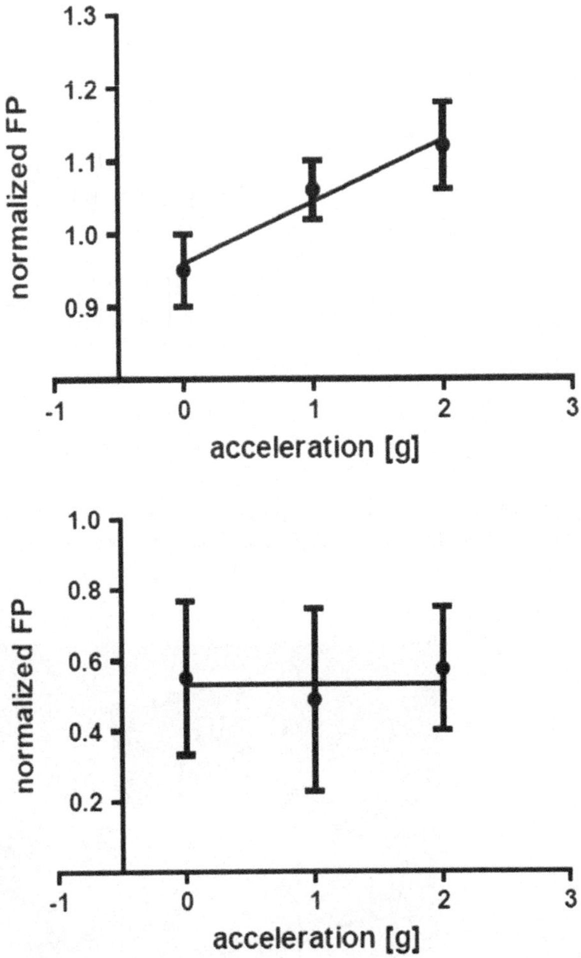

Fig. 1.12 Fluidity dependence of membrane fluidity in plain lipid vesicles. Increasing FP depicts decreasing fluidity. In the fluid state of the membrane, fluidity increases toward microgravity, left part. In the phase transition range of a given lipid, lower part, no gravity dependence of fluidity was found [modified from Sieber et al. (2014)]

(Wiedemann et al. 2011). In experiments with real cells (insect embryonic cancer cells), no changes in cell size and geometry were detected within the experimentally given resolution. This might be due to the fact that the cytoskeleton of the cells is preventing such changes.

In a given lipid membrane, membrane fluidity is correlated to the lateral pressure of the membrane (which is directly related to the membrane fluidity). In case the lateral membrane pressure changes, due to the correlation between pressure and area in an open system, the area of the membrane also changes (see Fig. 1.2). By this, the size of the vesicles might change due to gravity changes.

Further experiments with plain lipid bilayers were done utilizing the planar lipid bilayer voltage-clamp technique at high membrane potential ($V > 100$ mV). Under

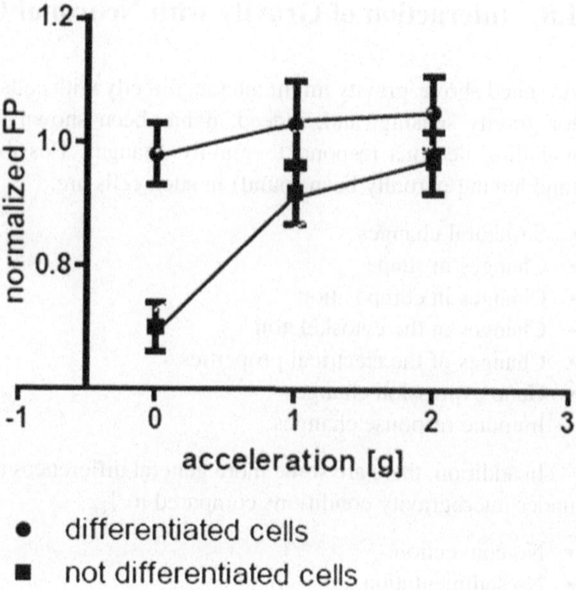

Fig. 1.13 Gravity dependence of membrane fluidity of neuronal cells. In both cell versions used, membrane fluidity increases with decreasing gravity levels [modified from Sieber et al. (2014)]

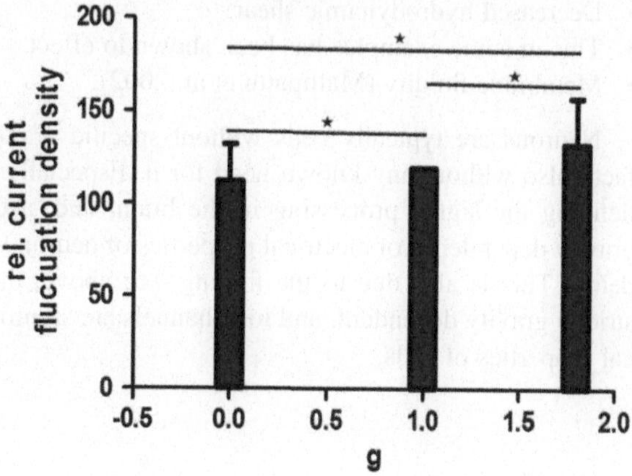

Fig. 1.14 Current fluctuation density at high membrane potential in asolectin bilayers at different gravity levels. Current fluctuation density induced by high potentials is reduced at microgravity [modified from Sieber et al. (2016)]

these conditions, current fluctuations can be induced in the plain lipid bilayer, which might be interpreted, for example, as lipid ion channels (i.e., (Heimburg 2010)). Such current fluctuations can be measured; the integral current density fluctuation in a bilayer measured in a parabolic flight mission is shown in Fig. 1.14. It decreases toward lower gravity levels.

1.8 Interaction of Gravity with Neuronal Cells

As stated above, gravity might interact directly with cells having no explicit structure for gravity sensing, and, indeed, it has been shown that a variety of such cells including neurons respond to gravity changes. Possible changes to be expected (and having partially been found) in such cells are:

- Structural changes
- Changes in shape
- Changes in composition
- Changes in the cytoskeleton
- Changes of the electrical properties
- Gene expression changes
- Immune response changes

In addition, there are some more general differences relevant for cells to be found under microgravity conditions compared to 1g:

- No convection.
- No sedimentation.
- Access to nutrients becomes diffusion limited.
- Waste dissipation becomes diffusion limited.
- Decreased hydrodynamic shear.
- This in some examples has been shown to effect.
- Membrane fluidity (Mallipattu et al. 2002).

Neurons are typically cells without specific gravity-sensitive structure and de facto also without any known need for it. Especially their electrical properties are defining the signal processing in the brain, and, accordingly, the clearly proven gravity dependence of electrical properties of neuron will be discussed here in more detail. This is also due to the finding, see above, that ion-channel properties are strictly gravity dependent, and ion channels are significantly involved in the electrical properties of cells.

1.9 Membrane Potential

One basic property of any cell is its resting membrane potential defined by the ion gradients and the selective membrane permeability given by ion channels, especially potassium channels. The ion gradients are established mainly by pumps consuming metabolic energy (ATP). The Goldman equation (or in simplified situations the Nernst equation) can be used to calculate the membrane potential. This membrane potential usually is measured by electrophysiological techniques utilizing, i.e., glass microelectrodes with electrometer amplifiers. Meanwhile, a variety of

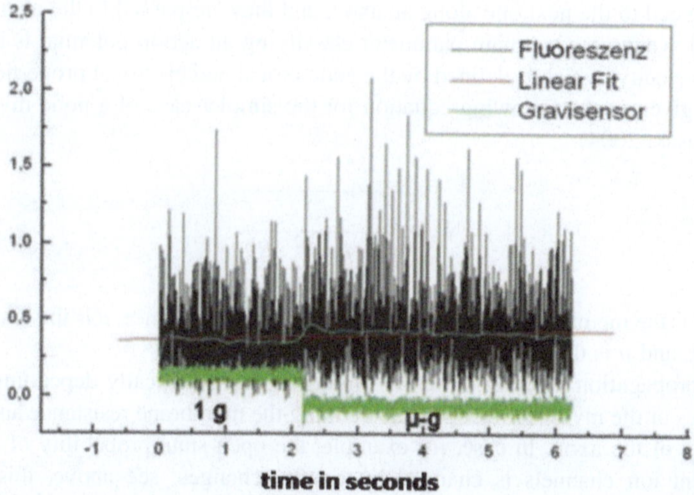

Fig. 1.15 Membrane potential changes in spherical SF 21 insect cells measured with the dye D-4-ANNEPS. The membrane potential becomes slightly more positive, indicating a depolarization of the cell membranes [modified from Wiedemann et al. (2011)]

potential-dependent fluorescent dyes is available, and optical techniques have entered the field of membrane potential measurement.

In a variety of different cells, membrane potential changes as a consequence of altered gravity have been reported (i.e., Wiedemann et al. 2011). As an example, in Fig. 1.15, the response of the membrane potential of an embryonic insect cancer cell to a gravity change is shown, as has been measured in an optical drop-tower experiment. A small membrane depolarization of an embryonic insect cancer was found as a consequence of the exposure of the cell to microgravity measured by the fluorescent potential-sensitive dye D-4-ANNEPS.

Membrane potential changes due to gravity changes have also been published for other unicellular organisms including neuron-like cells (Kohn 2013). These results were not completely systematic; sometimes, a hyperpolarization due to microgravity exposure was mentioned. In any case, in a neuron, changes in the membrane resting potential will result among others in differences in action potential initiation and propagation.

1.10 Action Potentials

Neurons in neuronal tissue process all incoming signals by integration, which leads to a specific membrane potential. In case the membrane is sufficiently depolarized relative to the resting potential of the cell, an action potential (AP) is generated in the cell. Neurons communicate by such action potentials, which are propagating from the firing cell to the next one along an axon, and they are passed to the next cell via chemical synapses. One main parameter classifying an action potential is its propagation velocity, which is defined by the geometrical and electrical properties of the axon as given in the following equation for the simpler case of a none myelinated axon (Tasaki 2004):

$$v = \sqrt{\frac{d}{8 \cdot \rho \cdot C^2 \cdot R^*}}$$

where C is the membrane capacity, R is the membrane resistance, d is the diameter of the axon, and p is the resistivity of the axoplasm.

The propagation velocity of an AP among others is critically depending on the properties of the involved ion channels defining the membrane resistance and on the geometry of the axon. In case, for example, the open-state probability of voltage-dependent ion channels is changed by gravity changes, see above, this should directly result in changes of the propagation velocity of the AP. Of course, especially voltage-gated sodium channels are relevant here, but for these up to now, no single-channel data under microgravity is available. However, when assigning the fact from above that some ion channels close upon microgravity to the axon membrane ion channels, the R in the equation would become bigger and the velocity would slow down, just consistent with our data. Additionally, in case a membrane becomes more fluid, it becomes thinner due to the higher mobility of the fatty acid chains of the lipids; thus, the capacity of the membrane becomes higher as it is

$$C = \varepsilon \times \varepsilon_0 \times A/d$$

where C is the capacity, ε the dielectrical constant of the medium (about 2 for lipids), ε_0 the general dielectrical constant, A the area, and d the thickness of the membrane. In addition, the axonal conductivity in case of a massive ion influx would increase, again in consistence with the other data.

Experiments concerning the question of action potential propagation velocity under microgravity using muscle APs in EMG and related studies have been done already as early as in the ninetieth of the last century (Layne and Spooner 1990) and in the following years (Rüegg et al. 2000). These experiments among others clearly showed the gravity dependence of muscle action potential propagation velocity (Pandis et al. 2009).

Also, in plants using *Dionaea muscipula* (Pandolfi et al. 2014), parabolic flight experiments were performed to investigate whether the electrical signaling of trap closure of this plant is depending on gravity, and it was suggested from the results that possibly the generation or propagation of the plant action potential was directly depending on gravity.

More detailed experiments were then done with isolated axons from rat (myelinated axon) and intact earthworm (good model system for none-myelinated axons), which demonstrated a basic dependence of the propagation velocity of action

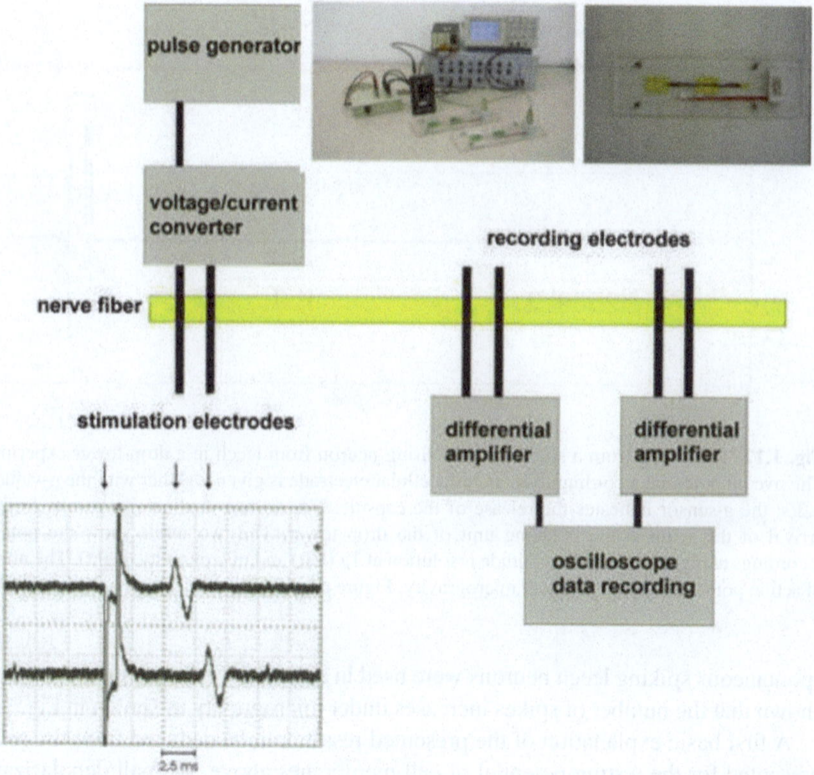

Fig. 1.16 In the experiment shown, an earthworm (or alternatively the rat ischiadicus) was prepared and placed in a proper chamber, here shown for the earthworm experiment. To this chamber, measuring electrodes (Ag/AgCl) were connected as shown in the sketch. Current pulses inducing membrane depolarization and thus initiating action potentials were applied to the fiber/ischiadicus. In the recording given at the lower left side from an earthworm, the stimulus artifact can be seen in both channels, followed by the recording of the propagating action potential. The propagation velocity was calculated from the time difference at the second recording electrode sets and the distance between these electrodes. Figure modified from Meissner and Hanke (2005)

potentials on gravity. A sketch of a typical experimental setup for a parabolic flight experiments with earthworm is shown in Fig. 1.16, together with an experimental trace.

By such experiments, it was clearly shown that APs propagate slower in microgravity (Wiedemann et al. 2011).

The second question related to the gravity dependence of AP properties is in how far their initiation is modified. As shown above, the membrane resting potential is gravity dependent. The relative value of the membrane potential compared to the threshold for action potential initiation is defining how easy an action potential can be triggered. To answer the question whether AP onset is gravity dependent,

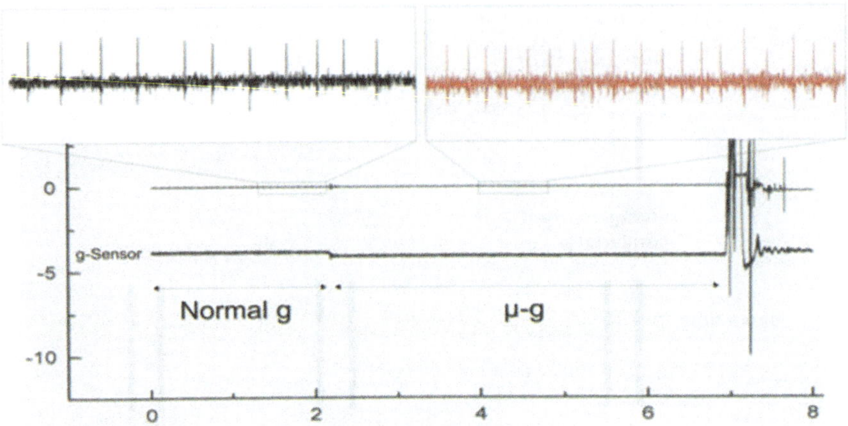

Fig. 1.17 Recording from a spontaneously firing neuron from leech in a drop-tower experiment. The overall potential recording from an intracellular electrode is given together with the g-value. At 2.2 s, the g-sensor indicates the release of the capsule. The artifact at about 7 s results from the arrival of the setup at the breaking unit of the drop tower. The two insets show the potential recordings at higher time and amplitude resolution at 1g (left) and microgravity (right). The number of action potentials increases under microgravity. Figure modified from Meissner and Hanke (2005)

spontaneous spiking leech neurons were used in a drop-tower experiment, and it was shown that the number of spikes increases under microgravity as shown in Fig. 1.17.

A first basic explanation of the presented results can be deduced from the results presented for the resting potential of cell membranes above. A small depolarization will lower the level for the initiation of an AP, and thus in the given experiment, the number of spontaneous APs will become bigger.

Related to the question of AP onset is the latency, the period between a proper stimulus and the onset of the AP. The shorter the latency is the easier is an AP to elicit. In the simplest case, this is analogue to the amount of membrane depolarization. Unfortunately, the terminus latency is often used more generally in physiological experiments, for example, describing the response of a muscle to an external stimulus. In such cases, not only the before defined latency but also AP propagation velocity and the behaving of the involved synapses is involved. Thus, it is difficult to compare the results in the literature.

It is important to mention here that there is no conflict in the statements that first, action potentials propagate slower on axons in microgravity and, second, are easier to elicit. In the first case, we are talking about a propagating wave; in the second case, an equipotential situation in a cell is given without propagation.

Due to the idea to look at membranes and cells as excitable media, it is of interest to have a look at other systems, which are discussed under the same assumption. Such systems can be used as model systems in some cases, as they possibly easier to be used in microgravity experiments or deliver additional information of interest.

Two of these systems are of more specific interest, as to both the Hodgkin-Huxley formalism of action potentials has been adapted.

The Belousov-Zhabotinsky reaction (Belousov 1959; Zaikin and Zhabotinsky 1970) is a chemical reaction exhibiting oscillatory behavior as well as propagating waves in thin layers of fluid or gels. BZ waves propagate much slower (5 mm/min) than APs and their mechanism is completely different. They are reaction-diffusion waves based on coupled chemical reactions. Nevertheless, propagating waves in the BZ can be described using similar differential equations as being used to describe propagating action potentials (i.e., Murray 2002) or any other propagating wave in an excitable medium. They also show the same behavior as action potentials in some other aspects; they annihilate upon collision, and they have a defined absolute and relative refractory period. Finally, their propagation follows a nonlinear dispersion relation as usually found for waves propagating in excitable media. Indeed, it was found that the propagation velocity of BZ waves depends on gravity too, but they speed up under lower gravity.

The other system having some similarities to action potentials but also to the BZ reaction is the retinal spreading depression (Fernades de Lima et al. 2015; Martins-Ferreira and de Oliveira-Castro 1966), a reaction-diffusion wave traveling through retinal tissue with a speed of about 5 mm/min, thus, similar to the BZ reaction. Again, its speed and its initiation are gravity dependent but much more complicated than found for action potentials. This is most probably due to the fact that here we are looking at a wave in neuronal tissue with its highly complex structure.

1.11 Cytosolic Calcium Concentration

As has been shown above already, the membrane resting potential of cells in most cases under microgravity slightly depolarizes. Most cells including neurons have potential-sensitive calcium channels in their membranes, which open upon depolarization. These channels might change their open-state probability under microgravity. Usually a high calcium gradient exists over cell membranes. Taking these facts into account, changes in intracellular calcium concentration, possibly an increase of intracellular calcium concentration, can be expected under microgravity.

Accordingly, experiments have been done in parabolic flight using neuronal cells together labeled with the dye Fluo-3 AM, which is sensitive to the intracellular calcium concentration. The experimental setup had been built around a commercial 96-well multipurpose plate reader as previously shown already. The result of such a measurement is shown in Fig. 1.18 (Hauslage et al. 2016).

Similar results were found in clinorotation experiments using the same cells (Hauslage et al. 2016) and also in *Arabidopsis thaliana* cells (Hausmann et al. 2014). Opening of membrane potential-dependent calcium channels due to membrane depolarization easily can explain these findings.

Interestingly, in the experiments by Hauslage et al. but also in *Arabidopsis thaliana* cells, it was found that under hypergravity the intracellular calcium

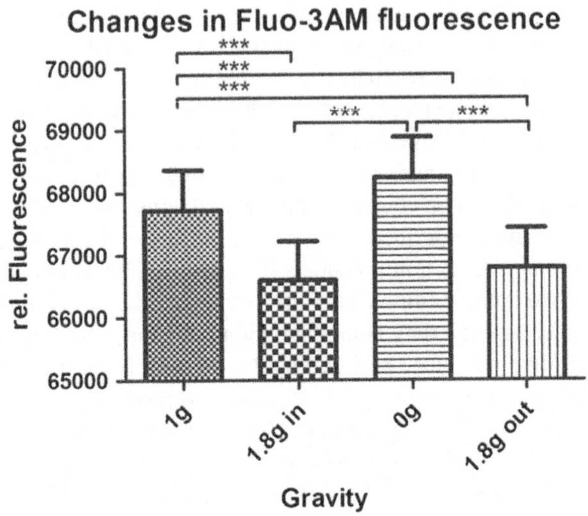

Fig. 1.18 Relative fluorescence of Fluo-3 AM incubated SH-SY5Y cells during parabolic flights. Intracellular calcium concentration increases significantly with decreasing gravity levels [modified from Hauslage et al. (2016)]

concentration decreases (Neef et al. 2015). This indicates that under hypergravity conditions calcium pumps must be activated further decreasing the intracellular calcium concentration. This would require a signal transduction cascade in the cells to be triggered by hypergravity activating ATP-driven or secondary energized calcium pumps.

1.12 Discussion and Consequences

The most general finding within the above presented results is the fact that the fluidity even of plain lipid membranes is gravity dependent (see also Klymchuk et al. 2006; Kordyum et al. 2015). The fluidity increases toward lower gravity values. To explain this finding mechanistically is not trivial. Our statement that biological systems behave like excitable media is no valid here, as plain liposomes with identical intra- and extra-liposomal aqueous solution are not far enough from equilibrium. Thus, the fluidity change cannot be energized by electrochemical gradients or metabolic energy. The most reasonable explanation to understand the effect is that it is thermodynamical driven. This should result in a small temperature shift of the system to lower values, which however, due to the high buffer capacity of the used setup, cannot be measured. Thus, to create an experiment, tackling this question will be of high biophysical importance. Alternatively, taking into account that mot membranes are not flat, but have certain ripple structures, interaction of gravity with this more complicated three-dimensional structure might be relevant.

Coming now to the consequences of membrane fluidity being gravity dependent, it is obvious that by this any cell will have a residual most probably small gravity dependence. This has been due of course from the first cell on earth to today, too,

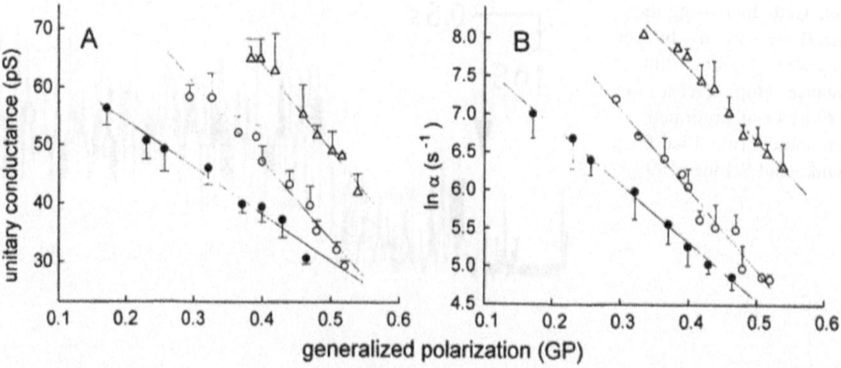

Fig. 1.19 Dependence of nicotinic AChR channel parameters on membrane fluidity as measured by fluorescence. The ion channel conductance increases toward higher fluidity, left side, and the closed-state probability of the channel decreases toward higher fluidity, this being in agreement with our findings for model pores polarization [modified from Zanello et al. (1996)]

and, thus, evolution has taken place under the permanent recognition of a stable gravity of about 9.8 m/s^2 (1g). Accordingly, some aspects of evolution should possibly be discussed taking this into account.

In higher organisms, having specific gravity sensors, the effect of membrane fluidity being gravity dependent might be ignorable, but not in single cells without such sensors as, for example, in neurons, but also in single-cell protists like *Euglena gracilis*. Especially according to the effect on neurons, a basic dependence of mental (brain) functions on gravity can no longer be excluded. Indeed, a variety of findings have been presented toward this direction (i.e., Wiedemann et al. 2011). However, the data about effects of gravity on brain function are complex, partially confusing, and even sometimes contradictory. In addition, they depend a lot on the set of the specific experiment. Thus, we will not go into details about gravity effects on the CNS here but will focus on membrane and cellular aspects.

It is furthermore accepted meanwhile that the function of membrane intrinsic proteins among others is depending on thermodynamical membrane parameters. Among these membrane parameters by sure is the membrane fluidity, and it has, for example, been shown that the parameters of the ion channel of the nicotinic acetylcholine receptor (nAChR) are directly depending on membrane fluidity as presented in the following figure from the literature.

From the results of Fig. 1.19, it can be concluded that the open-state probability of the nicotinic AChR decreases toward higher fluidity and thus toward lower gravity, which is consistent with our above presented results for the model pore alamethicin and the porin pore. The finding that the open-state probability of ion channels is gravity dependent, thus, just is a consequence of the change of membrane fluidity. Among others, the complete chemical synaptic transmission in neuronal systems and in the periphery should be gravity dependent.

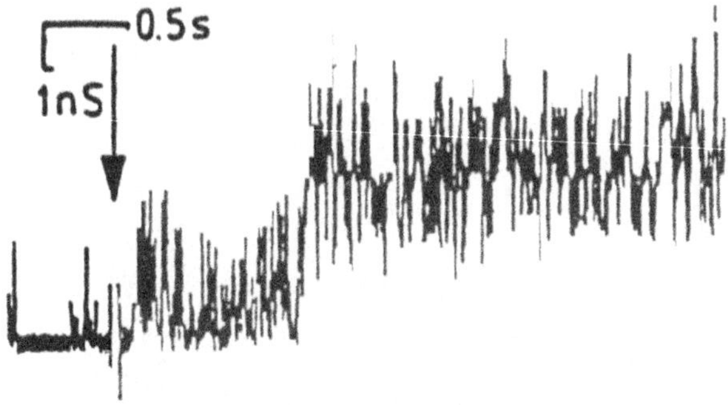

Fig. 1.20 Increasing the lateral pressure in a bilayer (arrow) in a double film-balance setup increases the activity of incorporated alamethicin [modified from Hanke and Schlue (1993)]

Furthermore, as the behavior of receptor proteins is membrane fluidity dependent, see the nAChR results shown above, the binding constants of ligands to other membrane receptors should also be gravity dependent. A major part of all relevant pharmacological drugs being used presently belongs to the substances binding to G-protein-coupled receptors (7-transmembrane-alpha-helix receptors, i.e., the muscarinic AChR). Accordingly, the activity of most of the relevant pharmacological substances might be gravity dependent. This must be considered seriously at least in later longer-lasting manned space missions. Finally, as stated already, the function of any other membrane protein could become gravity dependent, as, for example, ABC membrane transporters (Vaquer et al. 2014) or thyrotropin receptors (Albi et al. 2011).

Using alamethicin as sensor for lateral membrane pressure, it has been shown earlier (Hanke and Schlue 1993) that upon increasing the lateral pressure in a bilayer using a double film-balance setup, the activity of alamethicin also increases. This is consistent with higher activity at lower fluidity and at higher gravity values (Fig. 1.20).

Generalizing the finding that ion channels tend to close toward microgravity and using the equation for action potential as given above allows to state that the membrane resistance of axons will become higher and ,thus, action potentials will slow down under microgravity, as indeed has been shown in our experiments. Additionally, it was shown that the frequency of spontaneous action potentials in isolated neuron increases under microgravity, a finding being just the result of the depolarization of cells under microgravity. This depolarization in our interpretation comes from the closing of potassium channels under microgravity, which leads to a membrane resting potential depolarization as can be related from the Goldman equation. For this statement, we just have to use the additional dogma that the resting potential of cells almost exclusively is determined by the selective permeability of the membrane for potassium (as can be calculated by the Nernst equation).

It is important, however, to point out that spontaneous action potentials in isolated spiking neurons are not propagating, and, thus, their behavior is not directly related to the propagation of action potentials along axons.

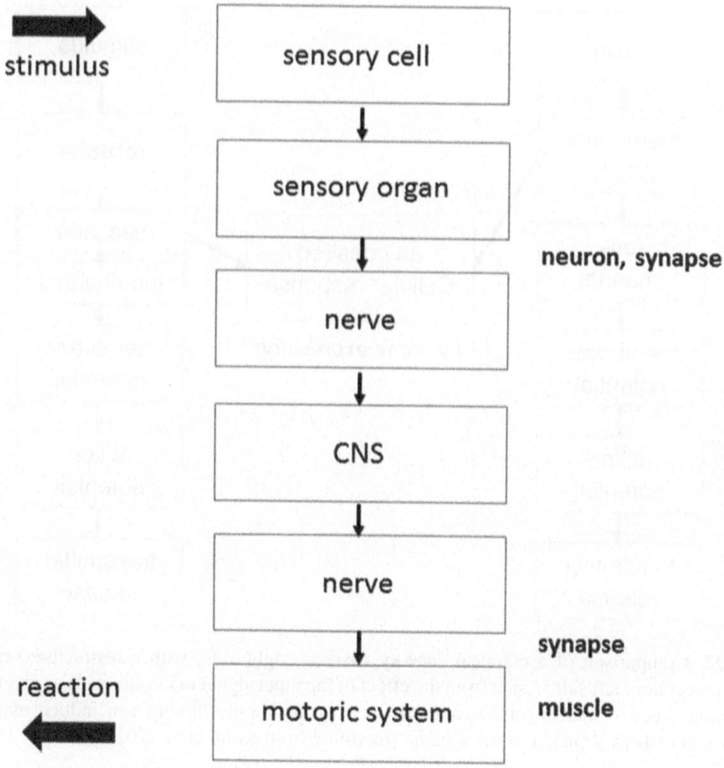

Fig. 1.21 Scheme of a sensory system

1.13 Modeling the Gravity Dependence of Neuronal Tissue

In sensory physiology, it is typically assumed that a proper sensor protein is given for the sensory stimulus in the plasma membrane of a given sensory cell. The sensory cell is embedded in a complete hierarchic system as shown in Fig. 1.21.

This receptor is located in the sensory cell membrane and might be an ion channel, directly activated by a stimulus, as, for example, pH-sensitive channels for acid recognition in the case of taste. It can also be a classical 7-transmembrane-helix receptor as given for olfactory stimuli or in vertebrate vision. Either such a receptor induces a direct membrane potential change or activates a second messenger cascade inducing again membrane potential changes but possibly also other actions.

Let us assume that it is allowed to make the membrane fluidity the sensory receptor for gravity, a physical property thus being a receptor for a physical stimulus. Then, the sensory transduction cascade can be redesigned in relation to classical sensory cascades as given in Fig. 1.22.

With these assumptions, we can now design a model from the above findings starting with microgravity as a stimulus being sensed by the receptor "membrane fluidity" and go forth to its final influence on coupled neurons as given in Fig. 1.23.

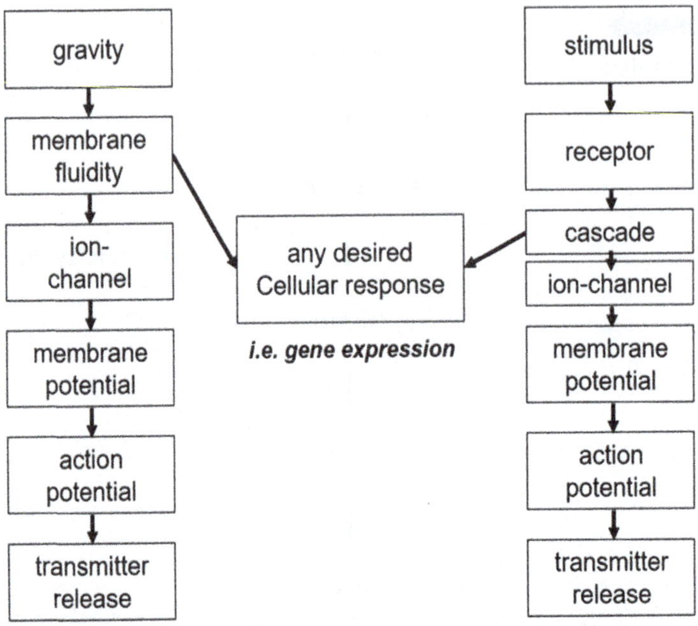

Fig. 1.22 Comparison of a classical sensory cascade, right side, with a restructured model of gravity perception, left side. Aside from the effect of the input signal to ion channels as given here as an example, second messenger cascades as well as membrane fluidity can induce many other processes in cells as depicted in the scheme [modified from Kohn et al. (2017)]

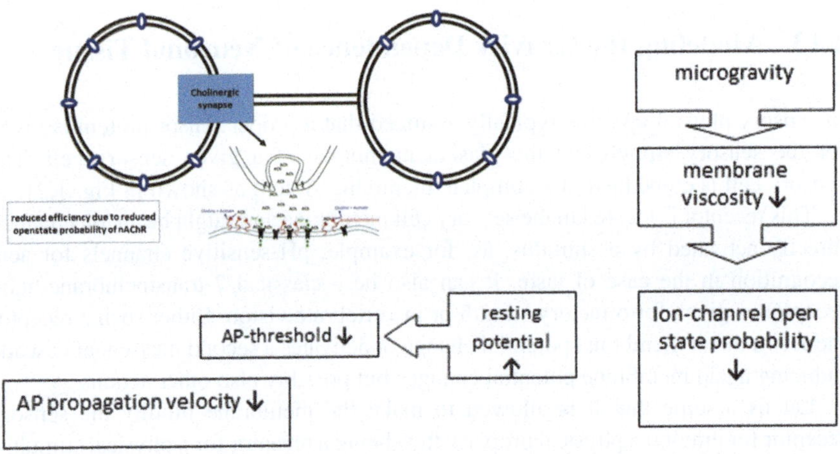

Fig. 1.23 The interaction of gravity with neuronal systems, from membrane fluidity to the coupling of neurons [modified from Kohn and Ritzmann (2017)]

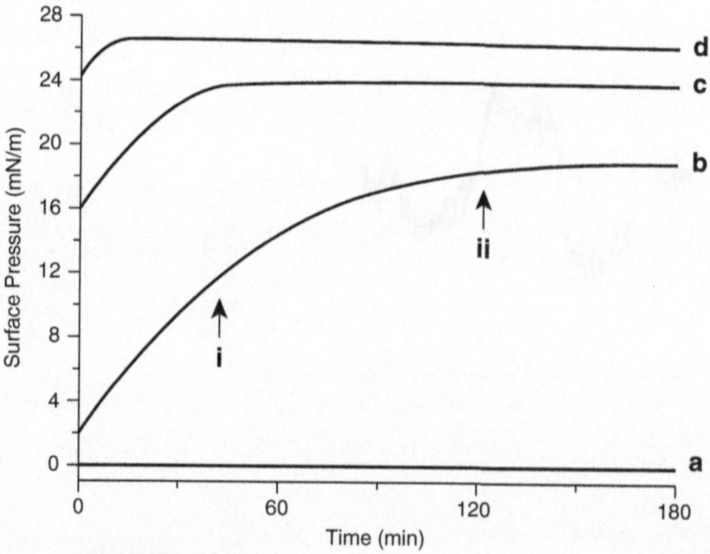

Fig. 1.24 Dependence of alamethicin incorporation into monolayers of different starting pressure [modified from Volinsky et al. (2004)]

1.14 Space Pharmacology

There is another major consequence of membrane fluidity changes due to gravity changes, as already shortly mentioned, related to later manned space missions, which is related to the relevance of the above shown findings to pharmacology. It is well known that the incorporation of hydrophobic and amphiphilic substances into membranes is clearly depending on lateral membrane pressure and thus on fluidity. A significant number of relevant pharmacologically relevant drugs including, for example, anesthetics, steroids, and others, belong to this group of substances. An example from the literature, again using alamethicin as sensor, is shown in Fig. 1.24. Here, in monolayers of different starting pressure, alamethicin was incorporated with clearly pressure-dependent kinetics (Volinsky et al. 2004).

From Fig. 1.24, a problem with the timescale becomes obvious. Usually the incorporation processes of hydrophobic and amphiphilic substances into membranes are relatively slow; thus, parabolic flight or other short-time microgravity platforms are not suitable for related experiments. Unfortunately, presently, no long timescale experiments concerning this question are available, but at least the data from one parabolic flight mission demonstrated that the kinetics of the incorporation of alamethicin into vesicle membranes is gravity dependent. Thus, it was tested whether the incorporation of hydrophobic substances into membranes might be gravity dependent, as it was known that the incorporation of substances into membranes is depending on membrane fluidity (see also later in the text).

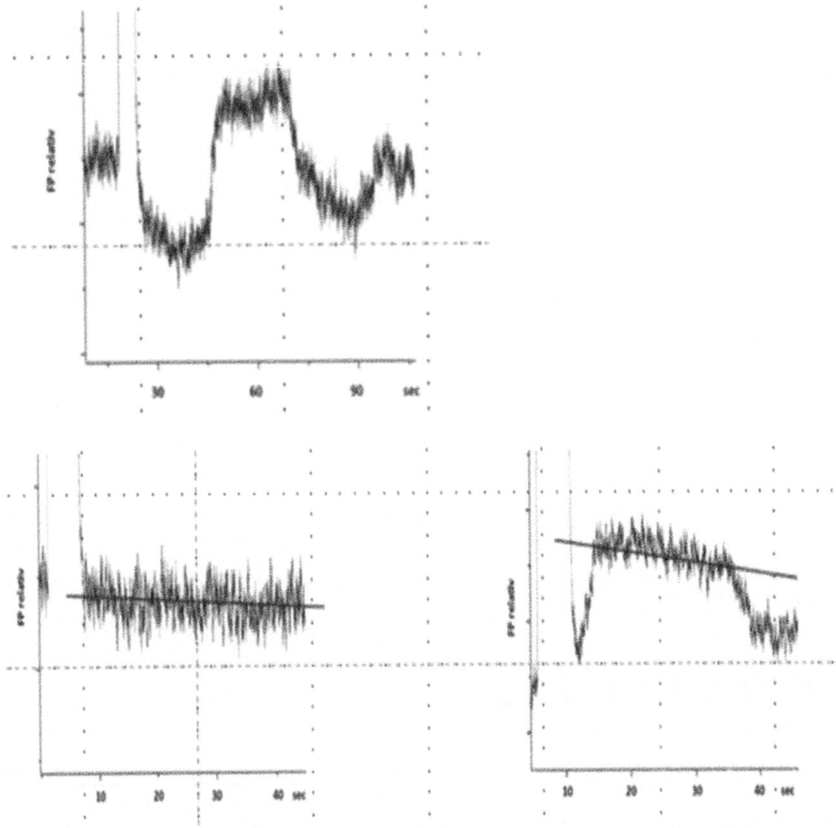

Fig. 1.25 Incorporation kinetics of alamethicin into vesicle membranes. Alamethicin was added to the vesicles at the beginning of each gravity phase in a modified stop-flow apparatus; this could be seen by the artifact at the beginning of each trace. The upper part shows the recording of a complete parabola, below, at an extended timescale is given for the 1g phase and for the microgravity phase. As can be seen, the slope of the incorporation is different

As a model system, plain lipid vesicles were used, to which alamethicin was added. Changes in lipid fluidity as measured by fluorescence polarization anisotropy were taken as a measure for alamethicin incorporation. The result is shown in Figs. 1.25 and 1.26.

The fluorescence polarization anisotropy technique was used for the experiments (data not yet published), and it was found that the incorporation kinetics of alamethicin into the vesicle membranes is changed under microgravity, being possibly a bit faster here.

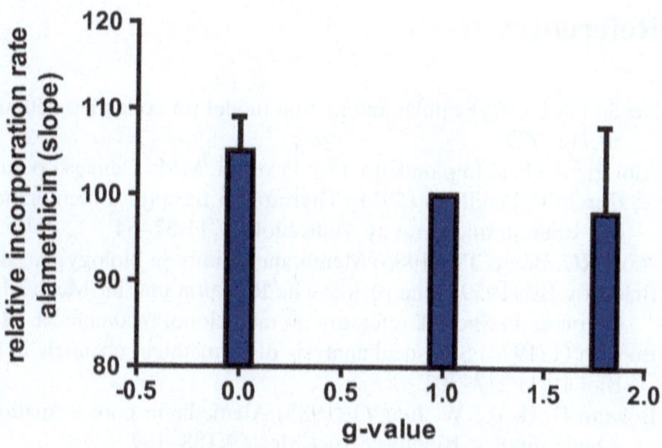

Fig. 1.26 Data evaluation from traces as shown in the previous figure. The incorporation of alamethicin into membranes is gravity dependent probably being slightly increased under microgravity; the results for the 1.8 g phases are scattering; thus, they are not discussed here

1.15 Outlook and Future Perspectives

Although a lot of information has been acquired about the effects of gravity on living systems, there is still a significant lack of important data. Systematic data about single-channel behavior under microgravity is still missing. Electrophysiological experiments under microgravity conditions up to now have been shown to be difficult, but possibly. Other techniques, mainly optical approaches, might deliver new possibilities for such measurements.

It is also obvious that the timescale of a variety of experiments is a problem together with the available microgravity platforms. In the above presented story, this is especially due for pharmacological experiments but also for many other questions. What is needed is microgravity in the range of one up to a few hours. The short-time platforms like sounding rockets cannot deliver much more than about 15 min of microgravity. By using orbital platforms, it would be technically feasible, but would be far too expensive, when only hours are needed. New approaches to solve this problem by sure would be welcome within the scientific community.

In case physiological functions are reacting to microgravity, it is obvious that they also should be reacting to hypergravity. Consequently, more experiments under high gravity should be done to expand our current knowledge. Such experiments can be done on centrifuges without relevant restrictions in time and in acceleration (gravity) level.

Finally, the problem of so-called simulated microgravity should be reconsidered, and it should be clarified to which extent and for which questions these techniques can be used. According to the presently available information in the literature, mainly clinostats might be useful here to some extent.

References

Acedo LA (2009) Cellular automation model for collective neural dynamics. Math Comput Model 50:717–725

Albi E, Ambesi-Impiombato FS, Peverini MM, Damaskopolus E, Fontanini E, Lazzarini R, Curcio F, Perella G (2011) Thyrotropin receptor a membrane interactions in FRTL-5 thyroid cell strain in microgravity. Astrobiology 11:57–64

Aloia RC, Boggs JM (1985) Membrane fluidity in biology. Academic Press, Orlando, FL

Belousov BP (1959) Eine periodische Reaktion und ihr Mechanismus (translated from Russian to German). In Sbornik referatov po radiacionoj medicine za. Moskau 147:145

Boheim G (1974) Statistical analysis of alamethicin channels in black lipid membranes. J Membr Biol 19(1):277–303

Boheim G, Hanke W, Jung G (1983) Alamethicin pore formation, voltage dependent flip-flop of a-helix dipoles. Biophys Struct Mech 9:188–197

Boheim G, Hanke W, Jung G (1984) The alamethicin pore is formed by a voltage-gated flip-flop of a-helix dipoles. In: Welter W et al (eds) Chemistry of peptides. W. de Gruyter, Berlin, pp 281–289

Cafiso DS (1994) Alamethicin: a peptide model for voltage gating and protein-membrane interactions. Annu Rev Biophys Biomol Struct 23:141–165

Dowhan W, Bogdanov M (2002) Functional roles of lipids in membranes. In: Vance DE, Vance JE (eds) Biochemistry of lipids, lipoproteins and membranes. Elsevier, Amsterdam, pp 1–35

Epstein IR, Pojman JA (1998) An introduction to nonlinear chemical dynamics. Oxford University Press, Oxford. ISBN 978-0195096705

Fernades de Lima VM, Piqueira JRC, Hanke W (2015) The tight coupling and non-linear relationship between the macroscopic electrical and optical concomitants of electrochemical CNS waves reflect the non-linear dynamics of neural glial propagation. Open J Biophys 5(1):11

Goldermann M, Hanke W (2001) Ion channels are sensitive to gravity changes. J Microgravity Sci Technol XIII/1:35–38

Goldman DE (1943) Potential, impedance, and rectification in membranes. J Gen Physiol 27:37–60

Häder D-P, Hemmersbach R, Lebert M (2005) Gravity and the behavior of unicellular organisms. Cambridge University Press, New York, NY

Hall JE, Vodyanoy I, Balasubramanian TM, Marshall GR (1984) Alamethicin. A rich model for channel behavior. Biophys J 45(1):233–247

Hamill OP, Neher E, Sakmann B, Sigworth FJ (1981) Improved patch-clamp techniques for high-resolution current recordings from cells and cell-free membrane patches. Pflügers Archiv 391:85–100

Hanke W (1995) Studies of the interaction of gravity with biological membranes using alamethicin doped planar lipid bilayers as a model system. Adv Space Res 6/7:143–150

Hanke W, Schlue W-R (1993) Planar lipid bilayer experiments: techniques and application. Academic Press, Oxford

Hauslage J, Albrecht M, Hanke L, Hemmersbach R, Koch C, Hanke W, Kohn PM (2016) Cytosolic calcium concentration changes in neuronal cells under clinorotation and in parabolic flight missions. Microgravity Sci Technol 28(6):633–638. https://doi.org/10.1007/s12217-016-9520-y

Hausmann N, Fengler S, Henning A, Franz-Wachtel M, Hampp R, Neef M (2014) Cytosolic calcium, hydrogen peroxide and related gene expression and protein modulation in *Arabidopsis thaliana* cell cultures respond immediately to altered gravitation: parabolic flight data. Plant Biol 16(Suppl 1):120–128

Heimburg T (2010) Lipid ion channels. Biophys Chem 150:2–22

Klinke N, Goldermann M, Hanke W (2000) The properties of alamethicin incorporated into planar lipid bilayers under the influence of microgravity. Acta Astron 47:771–773

Klymchuk DO, Baranenko VV, Vorobyova TV, Dubovoy VD (2006) Fluidity of pea root plasma membranes under altered gravity. http://adsabs.harvard.edu/abs/2004cosp.35.1356K

Kohn F (2010) Patch clamp experiments with human neuron-like cells under different gravity conditions. PhD thesis. University of Hohenheim, Stuttgart, Germany
Kohn F (2013) High throughput fluorescent screening of membrane potential and intracellular calcium concentration under variable gravity conditions. Microgravity Sci Technol 25(2):113–120
Kohn FPM, Ritzmann R (2017) Gravity and neuronal adaptation, in vitro and in vivo-from neuronal cells up to neuromuscular responses: a first model. Eur Biophys J. https://doi.org/10.1007/s00249-017-1233-7
Kohn F, Hauslage J, Hanke W (2017) Membrane fluidity changes, a basic mechanism of interaction of gravity with cells? Microgravity Sci Technol. https://doi.org/10.1007/s12217-017-9552-y
Kordyum EL, Neduhka OM, Grakhov VP, Melnik AK, Vorbyova TM, Klimeko OM, Zhupanov IV (2015) Study of the influence of simulated microgravity on the cytoplasmic membrane lipid bilayer of plant cells. Kosm Nauka Tehnol 21:40–47
Lacowicz JR (2006) Principles in fluorescence spectroscopy. Springer, New York, NY
Layne CS, Spooner BS (1990) EMG analysis of human postural during parabolic flight microgravity episodes. Aviat Space Environ Med 6:994–998
Leitgeb B, Szekeres A, Manczinger L, Vagvölgyi C, Kredics L (2007) The history of alamethicin: A review of the most extensively studied peptaibol. Chem Biodivers 4:1027–1051
Mallipattu SK, Haidekker MA, Von Dassow P, Latz M, Frangos J (2002) Evidence for shear-induced increase in membrane fluidity in the dinoflagellate *Lingulodinium polyedrum*. J Comp Physiol 188:409–416
Martins-Ferreira H, de Oliveira-Castro GD (1966) Light scattering changes accompanying spreading depression in isolated retina. J Neurophysiol 29:715–726
Meissner K, Hanke W (2002) Patch clamp experiments under microgravity. J Grav Physiol 9 (1):377–378
Meissner K, Hanke W (2005) Action potential properties are gravity dependent. Microgravity Sci Technol 17(2):38–43
Murray JD (2002) Mathematical biology. An introduction. Springer, New York
Neef M, Ecke M, Hampp R (2015) Real-time recording of cytosolic calcium levels in Arabidopsis thaliana cell cultures during parabolic flights. Microgravity Sci Technol 27:305–312
Novespace (2009) A300 Zero-G rules and guidelines, RG-2009-2
Pandis C, Metastasio A, Mastrandrea F (2009) Effects of gravity on ulnar nerve latency of activation. http://eea.spaceflight.esa.int/attachments/parabolicflights/ID49661e9539c5f.pdf
Pandolfi C, Masi E, Yoigt B, Mugnai S, Volkmann D, Mancuso S (2014) Gravity affects the closure of the trap in *Dionaea muscipula*. BioMed Res Int. https://doi.org/10.1155/2014/964203
Richard S, Henggeler D, Ille F, Vadrucci Beck S, Moeckli M, Forster IC, Franco-Obregón A, Egli M (2012) A semi-automated electrophysiology system for recording from Xenopus Oocytes under microgravity conditions. Microgravity Sci Technol 24(4):237–244. https://doi.org/10.1007/s12217-012-9307-8
Rüegg DG, Kakebeeke TH, Studer LM (2000) Einfluss der Schwerkraft auf die Fortleitungsgeschwindigkeit von Muskel-Aktionspotentialen. In: Kelle H, Sahm PR (eds) Bilanzsymposium Forschung unter Weltraumbedingungen. WPF, Aachen, pp 752–759
Sagués F, Epstein IR (2003) Nonlinear chemical dynamics. Dalton Trans 7:1201–1217
Sieber M, Hanke W, Kohn FPM (2014) Modification of membrane fluidity by gravity. Open J Biophys 4(4):105–111
Sieber M, Kaltenbach S, Hanke W, Kohn F (2016) Conductance and capacity of plain lipid membranes under conditions of variable gravity. J Biomed Sci Eng 9:361–366
Singer SJ, Nicholson GL (1972) The fluid mosaic model of the structure of cell membranes. Science 175(4023):720–731
Tabony J (2006) Microtubules viewed as molecular ant colonies. Biol Cell 98:603–617
Tasaki I (2004) On the conduction velocity of non-myelinated nerve fibers. J Integr Neurosci 3:115–124
Torchilin VP, Weissig V (2003) Liposomes: a practical approaches. Oxford University Press, New York

Vaquer S, Cuyas E, Rabadan A, Gonzale A, Fenollosa F, de la Torre R (2014) Active membrane drug transport in microgravity: a validation study using an ABC transporter model. F1000Research 3:201. https://doi.org/10.12688/f1000research.4909.1
Volinsky R, Kolusheva S, Berman A, Jelinek R (2004) Microscopic visualization of alamethicin incorporation into model membrane monolayers. Langmuir 20:11084–11091
Wiedemann M, Hanke W (2002) Gravity sensing in the central nervous system. J Grav Physiol 9 (1):43–44
Wiedemann M, Rahmann H, Hanke W (2003) Gravitational impact on ion channels incorporated into planar lipid bilayers. In: Tien HT, Ottova A (eds) Planar lipid bilayers and their applications. Elsevier Sciences, Amsterdam, pp 669–698
Wiedemann M, Kohn FPM, Rösner H, Hanke WRL (2011) Self-organization and pattern-formation in neuronal systems under conditions of variable gravity. Springer, Berlin. isbn:978-3-642-14471-4
Wolfram S (2002) A new kind of science. Wolfram Media, Champaign, IL
Woolley GA, Wallace BA (1992) Model ion channels: gramicidin and alamethicin. J Membr Biol 129(2):109–136
Zaikin AN, Zhabotinsky AM (1970) Concentration wave propagation in two-dimensional liquid-phase self-oscillating system. Nature 225:535–537
Zanello LP, Aztiria E, Antollini A, Barrantes FJ (1996) Nicotinic acetylcholine receptor channels are influenced by the physical state of their membrane environment. Biophys J 70:2155–2164
Zhao H, Lappalainen P (2012) A simple guide to biochemical approaches for analyzing protein-lipid interactions. Mol Biol Cell 23(15):2823–2830. https://doi.org/10.1091/mbc.e11-07-0645

Chapter 2
Interaction of Gravity with Cell Metabolism

Abstract Plants orient their organs, explore, and adapt to their environment mainly by sensing light and the direction of gravity. Some theories exist about gravity sensing including as a starting point the presence of dense sedimentable particles in specialized gravity-sensing cell types or protoplast-pressure phenomena inducing a cascade of biophysical and biochemical events that finally transform the directional information into gravioriented growth. However, apart from the directional information, plant cells show various more general gravity effects like changes in membrane-located processes and changes in gene expression, protein expression, and protein modulation as well as metabolic consequences in response to altered gravity conditions. In the following, mainly based on data from callus cultures of *A. thaliana*, we summarize the present knowledge in the field of gravity-affected cell metabolism, especially related to Ca^{2+} and hydrogen peroxide signaling.

Keywords *Arabidopsis thaliana* · Gene expression · Metabolic gravity response · Plants · Protein expression · Protein modulation · Secondary messengers

2.1 Introduction

Light and gravity are important environmental factors that control the development and growth of plants. With regard to gravity, this is quite obvious due to the spatial orientation of shoots, main roots, and first-order secondary roots. Investigations into the gravitational sensing of roots of higher plants give evidence for a cascade of biophysical and biochemical responses (Perera et al. 2001). For the initiation of such cascades, there are primarily two hypotheses. These are either based on the sedimentation of particles with high density such as starch-containing amyloplast statoliths (Sack 1991; Kiss 2000) or protoplast-pressure phenomena. The latter hypothesis suggests a perception of differences in pressure (reduced, increased) between the plasma membrane and the cell wall of opposing cell areas

© Springer Nature Switzerland AG 2018

W. Hanke et al., *Gravitational Biology II*, SpringerBriefs in Space Life Sciences,
https://doi.org/10.1007/978-3-030-00596-2_2

with the consequence of changed fluxes through mechano-sensitive ion channels (Staves 1997). Such a way of perception should function in any plant cell. It is also assumed that the actin cytoskeleton is included in response to altered gravity (Sievers 1991; Baluska and Hasenstein 1997; Rosen et al. 1999; Perbal and Driss-Ecole 2003; Soga et al. 2008). There is data, however, which indicates that at least in certain cell types, such as *Chara* rhizoids, actomyosin is more involved in fine tuning of responses than in gravity perception itself (Limbach et al. 2005; Braun and Limbach 2006). At the end of signal perception and transduction, there are metabolic responses, which help the cell to adjust to new gravitational steady states. These have been identified for primary, energy, and secondary metabolism (Obenland and Brown 1994; Hoffmann et al. 1996; Hampp et al. 1997).

Model systems which are primarily investigated are *Arabidopsis* and rice. For *Arabidopsis thaliana* with its small genome, large data banks exist, which support expression studies for both genes and proteins.

First reports about the long-term effect of microgravity (μg) on whole plants have been presented with material exposed on the International Space Station (ISS; Kiss et al. 2009; Pyle et al. 2001). A considerable number of gene transcripts were altered in amount including those involved in the jasmonic acid pathway. Downregulated genes included one coding for a peroxidase. Other ISS-related studies with dwarf wheat were not able to detect alterations in gene expression in 24-day-old leaves (Stutte et al. 2006). Changes in gene expression due to missing earth gravitation can be very fast. Callus cultures of *Arabidopsis thaliana* (A.t.) responded within minutes of microgravity as obtained by sounding rockets (Martzivanou et al. 2006). The impact of microgravity on gene expression results from perceived and transmitted signals (see above for gravity sensing). Independent of the external signal and its perception, environmental changes are transmitted by a transient increase of the intracellular calcium (Ca^{2+}) concentration. This has also been shown for changes in the gravity vector (Toyota et al. 2008).

Another trigger is the production of reactive oxygen species (ROS). They are no longer considered only as protectants against invading pathogens. ROS such as hydrogen peroxide have been shown to be also important in cellular signaling, and a "ROS gene network" has been suggested (Dodd et al. 2010; Neill et al. 2002; Miller et al. 2008). There are also reports about a close interrelationship between the hydrogen peroxide producing NADPH oxidase, hydrogen peroxide, and Ca^{2+} (Wong et al. 2007; Takeda et al. 2008).

Here, we mainly report about studies with callus cultures of *A. thaliana*. These deal with alterations in the pools of these secondary messengers, related gene and protein expression, as well as protein modification, and integrate metabolic alterations.

2.2 Methods

2.2.1 Opportunities for Exposure

Centrifugation (hypergravity) and clinorotation (2-D), or random positioning of samples (RPM), are well-established methods to investigate effects of gravistimulation in comparison with 1 *g*. In the case of clinorotation, the idea is that a constant stimulus acting similarly on all parts of a plant/tissue/cell could simulate microgravity. RPM treatments result in randomly disturbed cells, possibly prohibiting specific responses and thus simulating weightlessness. Magnetic levitation is a not so widely used technique, exploiting a strong magnetic field to counteract the earth's gravitational field and thus possibly simulating weightlessness.

2.2.1.1 Centrifugation

Application of *hypergravity* was done by centrifugation of Petri dishes (Fig. 2.1, Babbick et al. 2005; Martzivanou and Hampp 2003). After centrifugation, cells were scraped off and directly frozen in liquid nitrogen. The fluorometric analyses were performed with the centrifuge at the ZARM Institute in Bremen (http://www.zarm.uni-bremen.de/menu/facilities/centrifuge/; Neef et al. 2015).

Fig. 2.1 Petri dish centrifuge (University of Tübingen, Germany)

2.2.1.2 Clinorotation

Clinorotation was performed at 60 rpm with tubes having an internal diameter of 10 mm. This arrangement resulted in a maximal gravitational force of 0.0016 *g* (Hemmersbach, DLR, personal communication). Acceleration forces of 0.5 *g* and less were simulated with rotating Petri dishes (Fengler et al. 2015a, b).

2.2.1.3 Random Positioning

For random positioning, cultures were prepared on Petri dishes as described above for centrifugation. The dishes were then fixed in the center of the inner of the two connected frames. The frames were rotating in a random, autonomous way at an angular velocity of $60°s^{-1}$ (Walther et al. 1998). For details of the different procedures, see also Babbick et al. (2007).

2.2.1.4 Magnetic Levitation

A superconducting magnet with a closed-cycle liquid helium cooling system (Oxford Instruments, Abingdon, UK), capable of generating a very high magnetic field gradient, was used as another means to investigate the effects of simulated weightlessness on gene expression in *Arabidopsis* callus. For levitation, simulated weightlessness (sim 0 *g*) is defined as the condition, in which the magnetic force balances the weight of gravity of an object with certain caveats (Catherall et al. 2005). Cells (0.5 *g* f. wt.) were dispensed onto 3 ml aliquots of agar-solidified culture medium in 25 ml capacity, 2.7 cm diameter, screw-capped culture tubes (Sarstedt, Leicester, UK). After 7 days of culture, the tubes containing the samples were held at specific positions in the magnet bore in the dark using a purpose-built Perspex cylinder. The cells were exposed to 0 *g* and 1 *g* inside the magnet in the presence of a strong magnetic field. Cells at the 1 *g* position in the magnet were used as a control for the effect of the magnetic field alone (16.5 T). A constant temperature of 24 ± 1 °C was maintained in the magnet bore during the experimental period. Additional 1 *g* control cultures were placed in a temperature-controlled growth chamber in the dark outside the magnetic field.

2.2.1.5 Space Shuttle and Satellite

An extended period of microgravity was made available by the German-Chinese satellite mission Shenzhou 8. The respective flight and ground hardware (SIMBOX) were developed by Astrium GmbH (now Airbus D&S, Friedrichshafen, Germany) under a contract of the German Space Administration (DLR). The incubator contained a platform and a 1 *g* reference centrifuge. One out of two cultivation

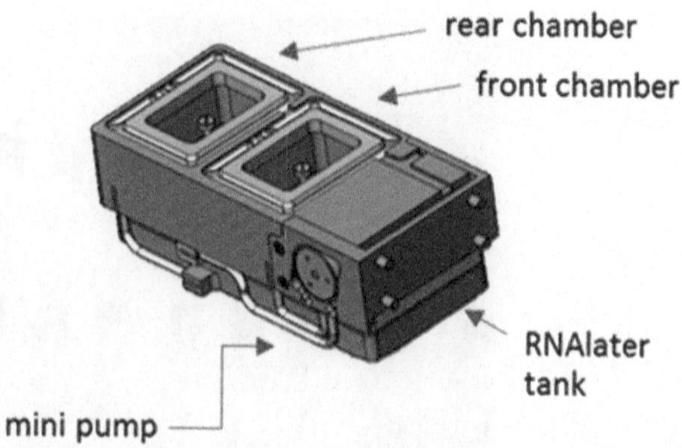

Fig. 2.2 Fully automated Type V plant cell cultivation unit (without cover) as used for Shenzhou 8

units (EUEs—experiment unique equipment) was fixed on the experimental platform and experienced microgravity. The second unit provided the in-flight reference by experiencing 1 *g* during the microgravity phase. Each module consisted of two cultivation chambers (Fig. 2.2).

In order to prevent anaerobiosis, samples were fixed after 5 days by the injection of the fixative RNAlater®. Finally, the samples were harvested and stored at 4 °C (Fengler et al. 2015a, b).

2.2.1.6 Sounding Rockets

This experimental setup was based on syringes. One of two 2 ml syringes was loaded with 90 mg callus culture and 30 μl liquid culture medium. A bead sitting in the tip of the syringe closed the vessel. Via an adaptor this syringe was connected to the other one, which contained the stop solution (1 ml RNAlater; Qiagen, Germany; Fig. 2.3). The syringes were placed in a module, sitting in the payload of the rocket. Total RNA from the cells was fixed at a defined time (i.e., 5 and 10 min μ*g*) by piston-actuated closure of the syringe, containing the stop solution.

For each time point, 8 incubation/quenching units were used in parallel. 1 *g* control samples were treated exactly the same way outside the payload. After recovery of the sample module by helicopter (within 1 h after launch), the RNAlater-treated samples were transferred into Eppendorf cups and kept between 4–8 °C until RNA extraction.

2.2.1.7 Parabolic Flights: Hyper-g, Partial g, and Microgravity

A typical parabola is illustrated in Fig. 2.4. The flights (Airbus A300, Novespace, France) took place between 6000 and 8500 m a.s.l. Parabolas consist of several phases: during ascend, the aircraft accelerates with 1.8 *g* (directed toward the ground

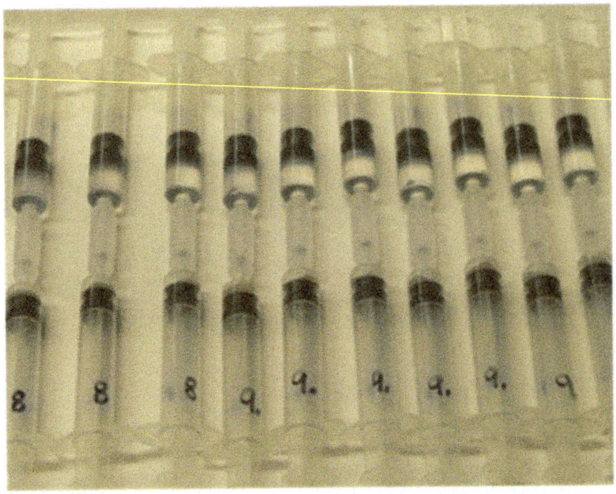

Fig. 2.3 Syringe-based quenching system. One syringe containing the cell culture is connected by a Teflon tube to a second syringe containing the quenching agent. The syringe with the quenching solution was sealed with a bead. When the cell culture syringe was activated at a given time, the bead was pushed out, and the quenching agent was injected into the culture tube (used for sounding rocket experiments)

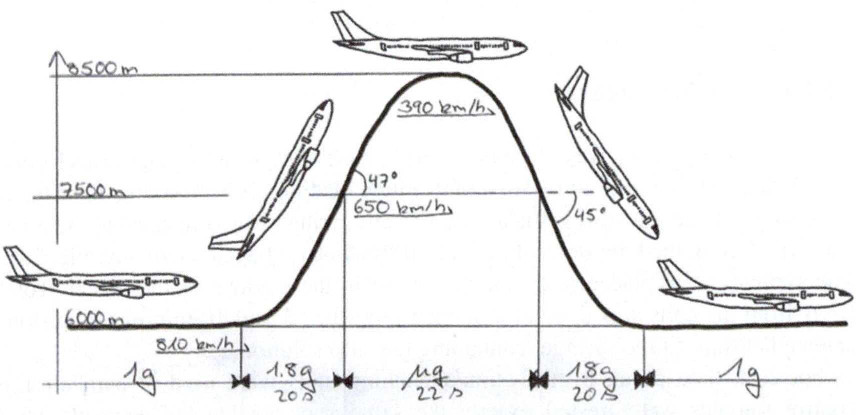

Fig. 2.4 Schematic representation of a parabolic flight maneuver. The numbers indicate manual sampling events for transcript and proteome analyses (modified after Hausmann et al. 2014)

floor) in an angle of about 47°. After about 20 s, the pilots strongly reduce the thrust, and the plane falls along the top of the parabola by its residual speed, with only some minor thrust to compensate for the air drag. During this flight phase, microgravity in a range of 10^{-2} g is provided for 22 s. Finally, within 20 s, the plane returns to a normal horizontal flight, causing an acceleration level of 1.8 g again. At 4 time points (indicated by numbers 1–4 in Fig. 2.4), samples were taken (e.g., for extraction of RNA and proteins) in cases, when no continuous measurements were made (e.g., fluorescence).

By adapting the flight angle and the overall velocity of the plane, the cells were additionally exposed to 0.38 and 0.16 *g* (Mars and Moon gravity during a Joint European Partial-G Parabolic Flight campaign).

2.2.2 Plant Material

Cell suspension cultures and calli were generated from leaves of *Arabidopsis thaliana* (cv. Columbia) plants grown under sterile conditions as detailed elsewhere (Barjaktarović et al. 2007). Calli with a diameter of about 1 mm were obtained after 1 week of growth and were used for the experiments.

2.2.3 Determination of Key Metabolites

Determination of ATP, ADP, and AMP from $HClO_4$ extracts was by luminescence according to Hampp et al. (1985, 1997). Quantitation of triphosphopyridine nucleotides was done by a cycling technique (BioVision, BioCat, Heidelberg, Germany; Hausmann et al. 2014). The determination of fructose 2,6-bisphosphate ($F_{26}BP$) was carried out according to Steingraber et al. (1988).

2.2.4 Metabolic Labeling with (^{14}C)-Glucose

During the TEXUS 37 sounding rocket campaign, we had the chance to perform radioactive labeling experiments. For this purpose, cell suspensions were incubated at about 24 °C in one of three 2 ml syringes, which were connected to each other via a T-piece (Fig. 2.5). The incubation with (^{14}C)-labeled glucose was started by closing the respective syringe and subsequent injection of the solution (incubation medium containing 1 mM glucose + 0.2 MBq of ^{14}C; specific activity, 12 GBq mmol^{-1}) into the syringe containing the cell suspension.

Labeling was terminated by the injection of ethanol (final concentration approx. 70%) from the third syringe (Fig. 2.5). Separation of metabolites contained in 2 µl aliquots was by thin-layer chromatography (TLC) on cellulose plates (Merck, Darmstadt, Germany) according to Feige et al. (1969). The spots were made visible by exposure to X-ray film (Amersham, Braunschweig, Germany) for about 3 months

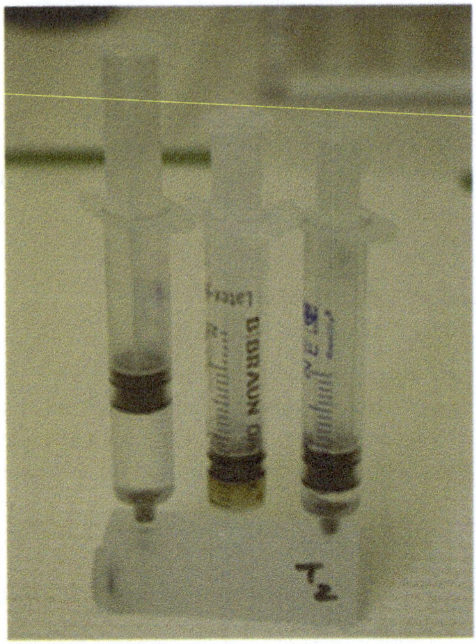

Fig. 2.5 Incubation of cell cultures for labeling experiments. Right (third syringe), labelled glucose; middle (second syringe), cell suspension; left (1st syringe), EtOH. The three syringes are connected via channels in a Perspex block and numbers 1 and 3 sealed by small beads. Syringes 1 and 3 are actuated in a time-dependent manner

at room temperature. In order to verify the different compounds, defined mixtures of the putative metabolites were run in parallel (+/− cell extracts).

2.2.5 *Real-Time Analysis of Ca^{2+} and Hydrogen Peroxide*

Seeds of *A. thaliana* plants expressing Yellow Cameleon were kindly provided by K. Schumacher/M. Krebs (University of Heidelberg, Germany). For construct details, see Allen et al. (1999) and Miyawaki et al. (1999). cHyPer cell culture lines, generated from shoots of 6-day-old transgenic seedlings (Costa et al. 2010), were obtained from A. Costa (University of Padua, Italy). Fluorescence of samples was recorded by means of a microplate reader (POLARstar Optima; BMG, Germany; Hausmann et al. 2014).

2.2.6 *Gene and Protein Expression*

Isolation of total RNA and protein, and microarray analysis (Affymetrix GeneChip) is detailed in Barjaktarović et al. (2007, 2009a, b), Neef et al. (2013a), and Hausmann et al. (2014).

2.3 Results

2.3.1 Metabolism

A consistent observation regarding metabolism of plants grown in microgravity or under horizontal clinorotation are changes in carbon metabolism. This can be a decrease (e.g., Obenland and Brown 1994; leaves exposed to microgravity), as well as an increase in starch accumulation (cell cultures; Wang et al. 2006).

Since the carbon metabolism is closely linked to energy metabolism, we assumed and finally could show that changes in gravity have an impact on the cellular energy metabolism (Hampp et al. 1997).

First attempts to study metabolic responses to changes in gravity were made during preparation (sounding rocket experiments) and execution of the D2 mission (1993). In these experiments, we studied pool sizes of adenylates, pyridine nucleotides, and a metabolite regulator of glycolysis, $F_{26}BP$.

Exposure of cell (protoplast) suspensions to microgravity on a short-term basis (sounding rocket flights) resulted in changes in the cellular energy and redox state within a few minutes of exposure (Hampp et al. 1997); i.e., the ratio of ATP/ADP started to fluctuate considerably reaching values of up to 2 as compared to 1 in 1 *g* controls. The decrease in NAD^+ indicated an increase of the NADH pool, while the rise of $NADP^+$ could be related to a decrease of NADPH (this is especially interesting with regard to the stress-related formation of reactive oxygen species as is discussed later). Centrifugation experiments also showed an increased NADH/NAD ratio in A.t. callus cells exposed for 1 h to between 6 *g* and 9 *g* (Fig. 2.6). Finally, random positioning resulted in an extended increase in this ratio between approx. 1 and 16 h of exposure (Maier et al. 2003).

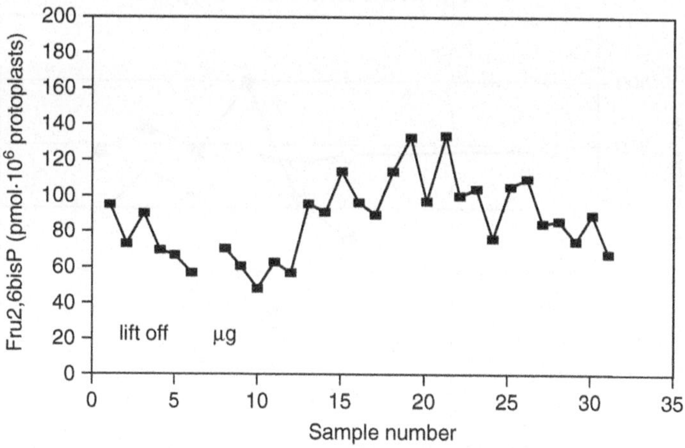

Fig. 2.6 Ratio of NADH/NAD in *A. thaliana* cell cultures during a sounding rocket flight (TEXUS). Samples were quenched every 20 s. The µ*g* phase ended around sample 26 (Hampp et al. 1992)

In addition, altered pool sizes of a regulator of plant cell glycolysis, fructose-2,6 bisphosphate (F26BP), were found. Under sounding rocket *microgravity*, this compound increased in pool size (Hampp et al. 1997; Fig. 2.6). As elevated levels of F26BP favor glycolysis, we assumed that this could be related to an enhanced use of intermediates of glycolysis for either respiratory ATP production or de novo synthesis of proteins. Another indication that changes in the effective gravitational field affect glycolysis comes from hyper-*g* (7 *g*) and random positioning experiments (Maier et al. 2003).

Here, increased ratios of the glycolysis intermediates, pyruvate/phosphoenolpyruvate (PEP), were found (Fig. 2.7). These suggest a piling up of pyruvate at an increased consumption of PEP. Together with an increased PEP carboxylase expression (PEPC; see below; Maier et al. 2003), this was taken as indicative for stress compensation (increased rates of anaplerosis related to protein synthesis).

The enzyme phosphoenolpyruvate carboxylase (PEPC) is a key component of anaplerosis. Due to a limited availability of flight opportunities, we thus focused on centrifugation experiments. *Hypergravity* resulted in a transient increase of PEPC activity after 1 h exposure to g-forces between 6 and 10 (Fig. 2.8).

This was obviously due to de novo synthesis, as both enzyme-specific protein (Western blot; Fig. 2.9) and the amount of the respective transcript increased (Maier et al. 2003; Fig. 2.10). Posttranslational modulation was not involved, as the activation (= phosphorylation) state of the enzyme was not altered. Pool sizes of pyruvate (pyr), phospho*enol*pyruvate (PEP), and the ratio of NADH/NAD suggested an inhibition of the pyruvate dehydrogenase complex in favor of conversion of PEP to oxaloacetate, which is the precursor for the synthesis of amino acids of the aspartate family.

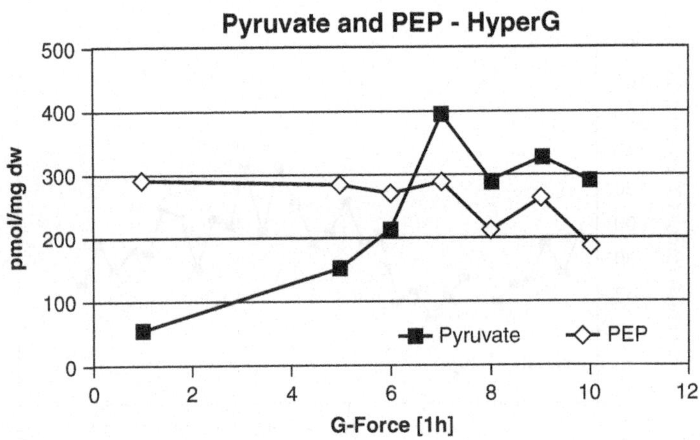

Fig. 2.7 Contents of pyruvate and phosphoenolpyruvate (PEP) in *A. thaliana* cell cultures which were treated for 1 h at different *g*-forces

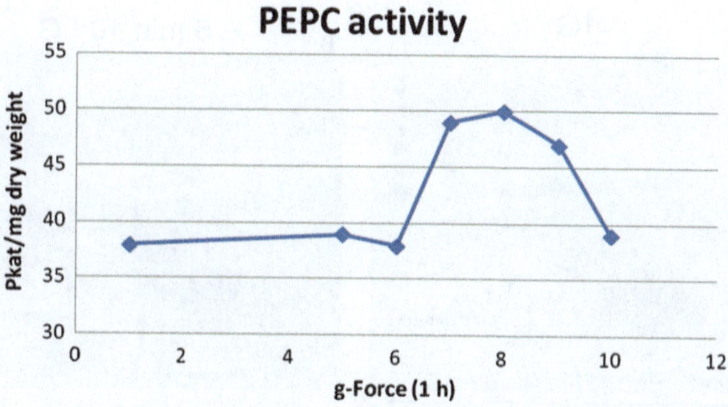

Fig. 2.8 Activity of phosphoenolpyruvate carboxylase (PEPC) from *A. thaliana* cell cultures upon centrifugation for 1 h at different *g*-forces

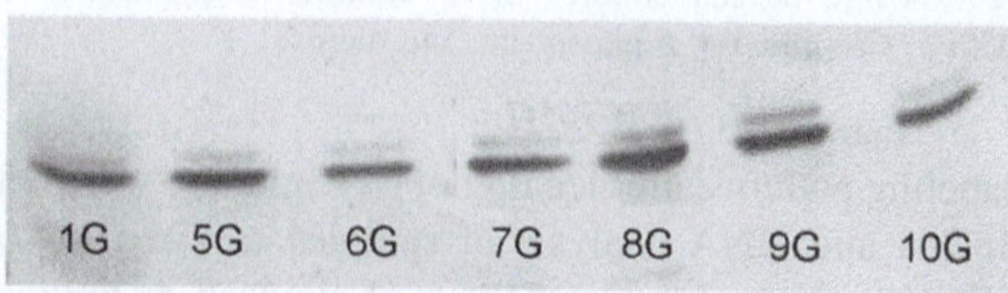

Fig. 2.9 Western blot with phosphoenolpyruvate carboxylase-specific antibodies. *A. thaliana* cell cultures were exposed for 1 h at different *g*-forces

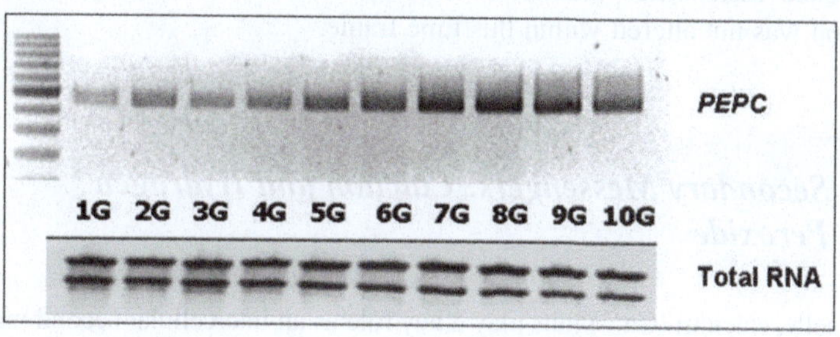

Fig. 2.10 Phosphoenolpyruvate carboxylase-transcript, detected by RT-PCR in *A. thaliana* cell cultures which were exposed for 1 h at different *g*-forces. Total RNA was used as a reference

Clinorotation of the cells also caused a transient increase in PEPC activity (between 30 min and 16 h), which was not mirrored by changes in the amounts of the respective transcript.

Under clinorotation, the enzyme is probably regulated by posttranslational modification. Independent of the way of regulation, pool sizes of pyr, PEP, and the ratio of NADH/NAD changed as shown for hypergravity conditions. This suggests a similar response but triggered in a different way.

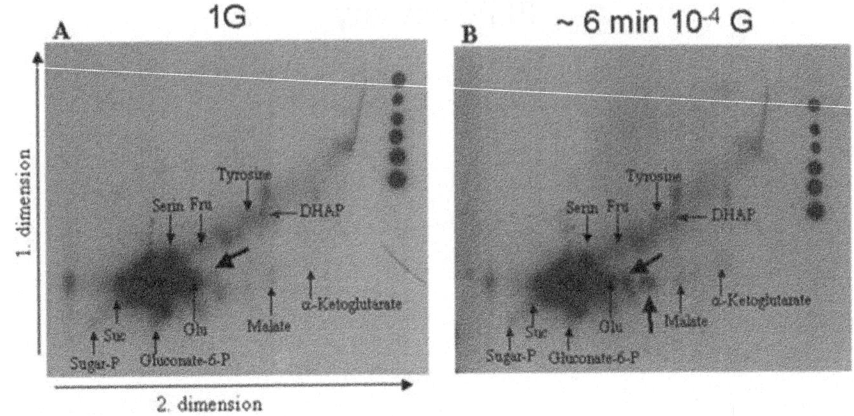

Fig. 2.11 Autoradiographs of the fixation pattern after incubation with (^{14}C)-labeled glucose of the 1 *g* control (**a**) and after 6 min of microgravity (TEXUS) (**b**). The arrows indicate three spots which are increased after exposure of cell cultures of *A. thaliana* to µ*g*. *DHAP* dihydroxyacetone phosphate; *Fru* fructose; *Glu* glucose; *P* phosphate; *Suc* sucrose

In addition, labeling patterns after feeding of (^{14}C)glucose were determined. Both labeling experiments and RNA analysis of samples exposed to sounding rocket microgravity (TEXUS) showed fast responses of exposed *A. thaliana* cell suspensions. With regard to labeling, we detected an increase in the abundance of three metabolites of the organic/amino acid fraction after 6 min of microgravity. The nature of these intermediates could not be resolved, because cochromatography with 53 reference metabolites did not reveal identical Rf values (Fig. 2.11). PEPC expression was not altered within this time frame.

2.3.2 *Secondary Messengers: Calcium and Hydrogen Peroxide*

In plant cells, calcium (Ca^{2+}) ions play a key role as an intercellular second messenger in response to a variety of environmental stimuli such as light, touch, pathogenic elicitors, plant hormones, high salinity, cold, and drought (Dodd et al. 2010; Sarwat et al. 2013). Spatially and temporally distinct changes of the cytosolic Ca^{2+} concentration are thought to constitute a unique signal for its inducing stimulus. Such stimulus-specific spikes in the cellular Ca^{2+} concentration are also known as Ca^{2+} signatures and result usually from two opposing responses: Ca^{2+} influx through channels or Ca^{2+} efflux through pumps (Tuteja and Mahajan 2007). Up to now, three different types of cytosolic Ca^{2+} signatures are described in the literature: oscillatory (e.g., after cold, ABA, or nod-factor treatment); biphasic, with a large, rapid first increase and a second smaller but sustained increase (e.g., in response to bending); and monophasic with a single cytosolic Ca^{2+} increase (e.g., due to red/blue light, salinity, heat, and touch; Batistic and Kudla 2012). Also, gravity sensing is thought to be transmitted by changes in the concentration of cytosolic Ca^{2+} (Poovaiah et al. 1987, 2002; Plieth and Trewavas 2002; Gehring et al. 1990; Gilroy et al. 1993; Fasano et al. 2002; Correll et al. 2013). More recently, visualization and analysis of intracellular Ca^{2+} dynamics in living plants were revolutionized through the introduction of genetically encoded Ca^{2+} sensors (e.g., aequorin or GFP-based calcium probes like YC3.6; Knight et al. 1997; Allen et al. 1999; Nagai et al. 2004). Toyota and coworkers used *A. thaliana* seedlings expressing cytoplasma-targeted apoaequorin for their analyses and described a biphasic cytosolic Ca^{2+} increase during parabolic flights, which included a plant rotation in addition to the gravistimulation (Toyota et al. 2013; Toyota and Gilroy 2013). Without rotation, a delayed and sustained monophasic cytosolic Ca^{2+} increase was observed (Toyota et al. 2013). For our analyses, we used a YC3.6 construct, which possesses reduced pH sensitivity and, owing to the UBQ10 promoter, exhibits a high rate of expression in *A. thaliana* (Krebs et al. 2012; Behera et al. 2013). With this construct we found cytosolic Ca^{2+} fluctuations which run parallel to gravitational changes (hypergravity and microgravity) during parabolic flights (Fig. 2.12).

In order to get an idea about the source of the μ*g*-related increase of cytosolic Ca^{2+}, we used a set of known and established Ca^{2+} inhibitors/antagonists to affect either the Ca^{2+} influx into or the Ca^{2+} efflux out of the cell. The data show a fast rise in cytosolic Ca^{2+} upon the onset of microgravity and indicate that cytosolic Ca^{2+} is mainly of extracellular origin. In contrast, exposure to hypergravity resulted in decline of the cytosolic amount of Ca^{2+} starting from between 3 *g* and 4 *g* (Fig. 2.13; Neef et al. 2016).

Another trigger for the transmission of environmental changes is the production of reactive oxygen species (ROS). They are no longer considered only as protectants against invading pathogens. ROS such as hydrogen peroxide have been shown to be

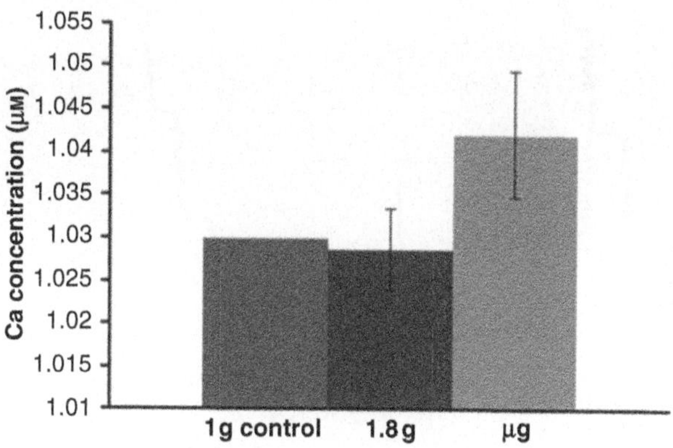

Fig. 2.12 Intracellular Ca^{2+} concentration depending on gravity. The values are means of 24 parabolas ±SD ($n = 24$). The control was set as 100% [1.03 μM; mean of all control readings (1 *g*)]. The differences between 1.8 *g* and microgravity, as well as between 1 *g* and microgravity, are highly significant according to Tukey's multiple comparison of means (***$P < 0.001$). There was no significant difference between in-flight 1 *g* and initial 1.8 *g* (Hausmann et al. 2014)

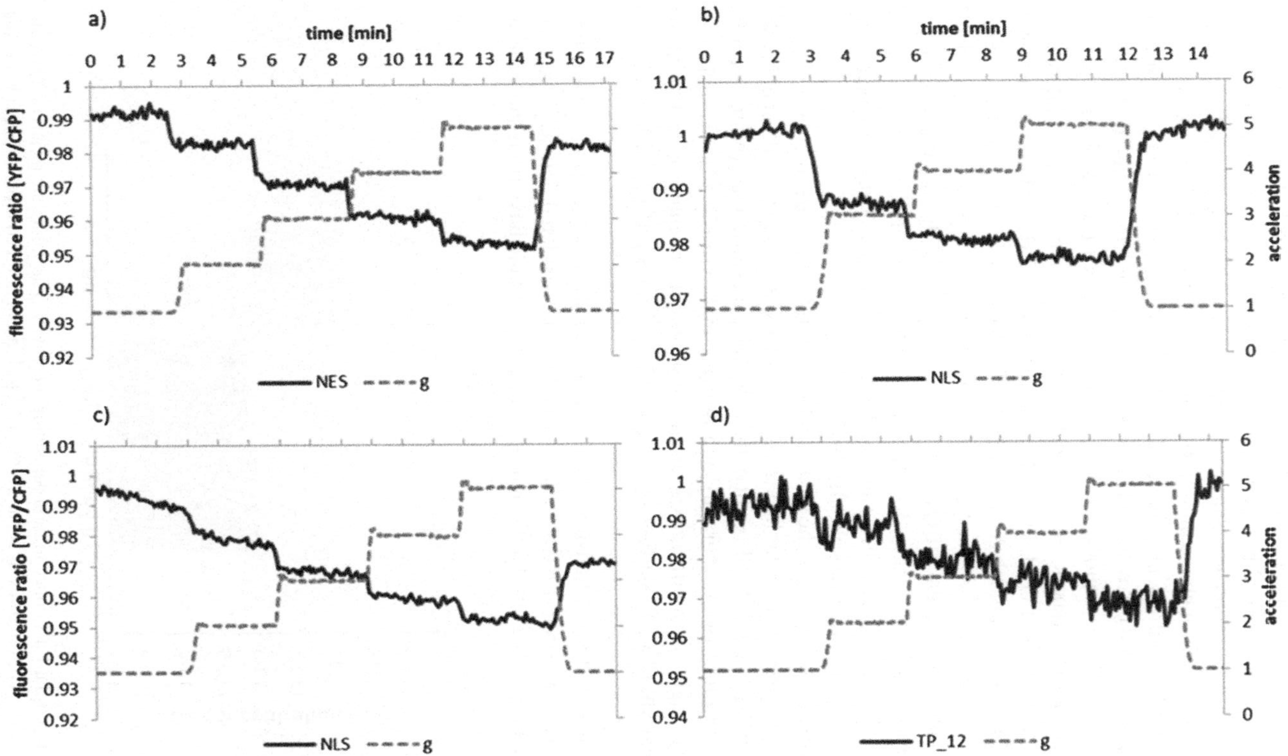

Fig. 2.13 Ca^{2+} transients in different intracellular compartments under hypergravity. Primary *y*-axis, fluorescence ratio (530/480 nm; black line); secondary *y*-axis, acceleration (*g* gray dotted line); *x*-axis, time (minutes); (**a**) NES (cytosol and nucleus), (**b**, **c**) NLS (nucleus), (**d**) TP12 (tonoplast) (Neef et al. 2016). NES, NLS, TP12: compartment-specific expression of the cameleon construct (Neef et al. 2015)

also important in cellular signaling, and a "ROS gene network" has been suggested (Dodd et al. 2010; Neill et al. 2002; Miller et al. 2008). There are also reports about a close interrelationship between the hydrogen peroxide producing NADPH oxidase, hydrogen peroxide, and Ca^{2+} (Wong et al. 2007; Takeda et al. 2008). For the determination of hydrogen peroxide, we employed cell cultures expressing the yellow fluorescent protein-based sensor HyPer (Belousov et al. 2006). This sensor is specific to this component of reactive oxygen species. For experimental details, see Neef et al. (2015).

Here, we also found an increase in this ROS component upon lack of gravity (Fig. 2.14). Altogether, we evaluated about 60 parabolas during independent flights in 2011 and 2012. The responses of the cell cultures were more variable for hydrogen peroxide than for Ca^{2+}. The high number of "non-responding" cultures (40%) could also be due to the lower sensitivity of the HyPer system.

Hydrogen peroxides can be produced by NADPH oxidase, located in the plasma membrane. This enzyme is activated by Ca^{2+} and releases hydrogen peroxide to the apoplast (Wong et al. 2007). It is fueled by intracellular NADPH, which probably is generated by glucose 6P dehydrogenase (oxidative pentose phosphate cycle). If indeed significant amounts of hydrogen peroxide are formed during onset of microgravity, pool sizes of NADPH should respond. We thus assayed the amounts of NADPH and NADP. In this case samples were taken at the points indicated in Fig. 2.4 and quenched in either acid (NADP) or base (NADPH). The calculated ratios are given in Fig. 2.15. At 1 *g* (before acceleration), the ratio is around 1.2. At the onset of μ*g*, the ratio drops (0.9) but rises during the μ*g* phase, reaching nearly 1.7 at the end. At the end of the second 1.8 *g* phase, the ratio comes back to control values. Especially the behavior during microgravity, which is opposite to that of hydrogen peroxide, indicates some interdependence: the onset of hydrogen peroxide

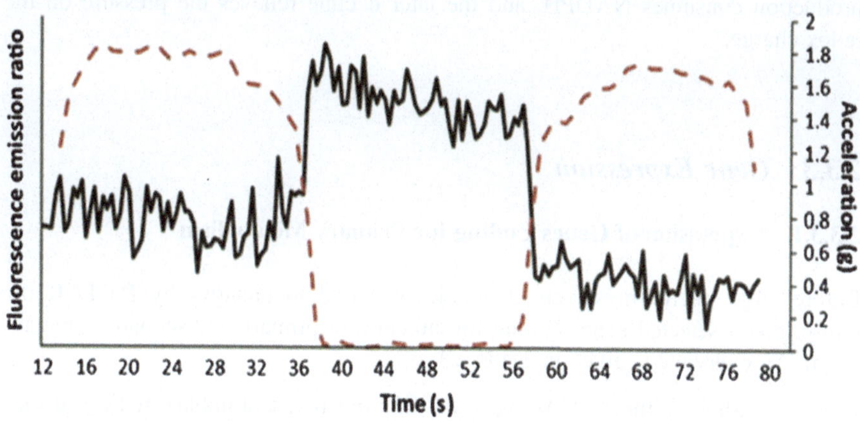

Fig. 2.14 Amounts of hydrogen peroxide, shown as fluorescence emission ratio during a parabola (excitation ratio 480:420 nm/emission 530 nm). An increase in the ratio indicates an increased H_2O_2 content. The graph presents the typical behavior of a responsive callus. The dotted line gives the respective kinetics of the acceleration, ranging from <0.01 *g* to about 1.8 *g* (Hausmann et al. 2014)

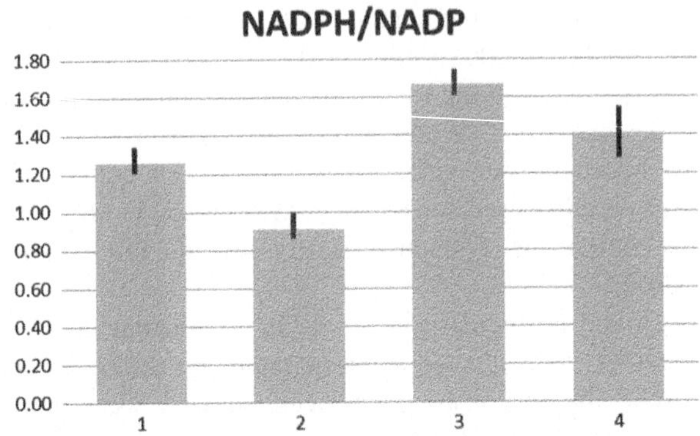

Fig. 2.15 The ratio of NADPH/NADP during parabolic flights. Numbers 1–4 represent the flight phases shown in Fig. 2.4. Bar: SD ($n = 7$) (modified after Hausmann et al. 2014)

Table 2.1 Expression of genes coding for primary metabolism

Protein	Metabolic function
Phosphoenolpyruvate carboxylase (PEPC)	Anaplerosis
∞-Amylase	Starch degradation
ADP-glucose pyrophosphorylase (LSU)	Starch synthesis
Fructose bisphosphatase (FBPase)	Gluconeogenesis
Glucose-6-phosphate dehydrogenase (G6PDH)	Ox. pentose phosphate cycle
Glyceraldehyde phosphate dehydrogenase (GAPDH)	Glycolysis
Phenylalanine ammonium lyase (PAL)	Secondary metabolism
Hydroxymethylglutaryl (HMG)-CoA reductase	Secondary metabolism

production consumes NADPH, and the later decline relieves the pressure on the redox charge.

2.3.3 *Gene Expression*

2.3.3.1 Expression of Genes Coding for Primary Metabolism

Before large microarrays were available, we tried to quantify by RT-PCR the transcripts of selected genes coding for enzymes of primary and secondary metabolism. An overview is given in Table 2.1.

Hyper-g Callus cultures of *Arabidopsis thaliana* (cv. Columbia) in Petri dishes were exposed to altered *g*-forces by centrifugation (1–10 *g*). Using semiquantitative RT-PCR, transcripts of genes coding for metabolic key enzymes were used to monitor threshold conditions for *g*-number (all) and time of exposure (β-amylase) which led to altered amounts of the gene product.

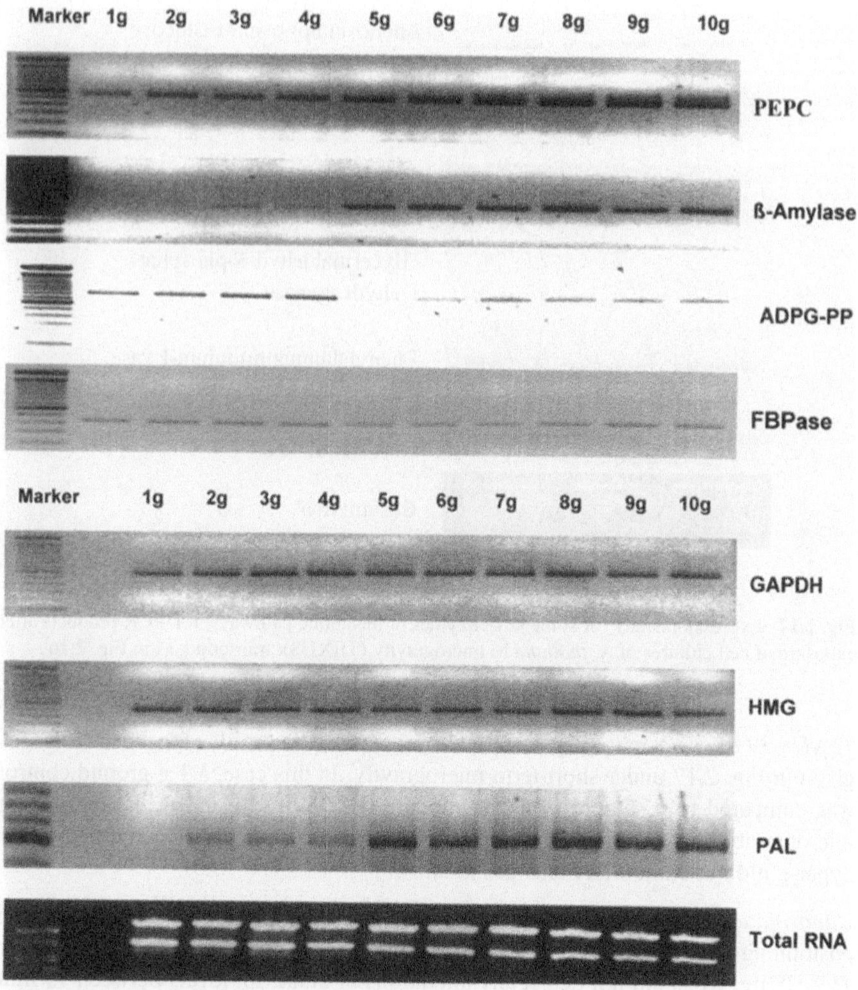

Fig. 2.16 Expression study for some key enzymes of metabolic pathways. RT-PCR products after separation on 2% agarose and ethidium bromide staining from cell cultures of *A. thaliana* exposed to g-forces between 1 and 10 for 60 min. As reference, the amount of total RNA is given for each sample. *ADPG-PP*, ADP-glucose pyrophosphorylase; *FBPase*, fructose-1,6-bisphosphatase; *GAPDH* glyceraldehyde-P dehydrogenase; *HMG* hydroxymethylglutaryl-CoA reductase; *PAL* phenylalanine ammonium lyase; *PEPC* phosphoenolpyruvate carboxylase

As reference total RNA was taken. Controls were samples exposed to 1 *g*. The data show that starting from 5 *g*, transcripts involved in anaplerosis, starch degradation, and gluconeogenesis were increased in amount, while those of starch synthesis were downregulated. With the exception of PAL, transcripts involved in glycolysis and secondary metabolism were not affected (Fig. 2.16). This probably indicates a shift from starch synthesis to starch degradation and increased rates of anaplerosis [PEPC, see above ("metabolism")].

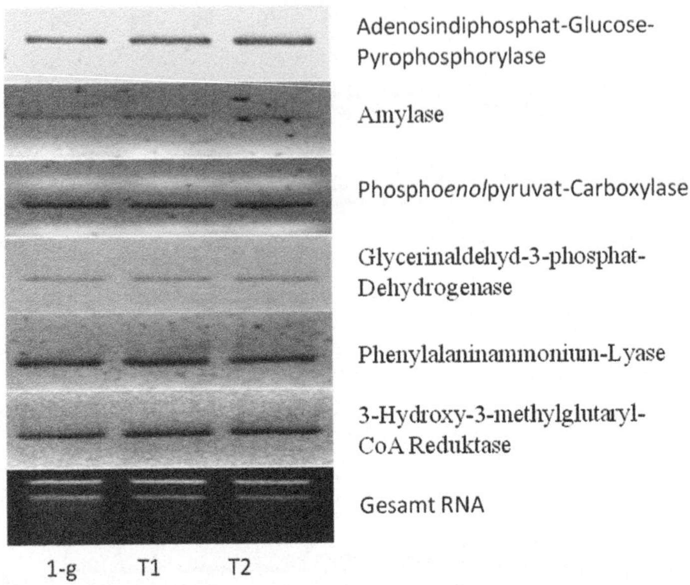

Fig. 2.17 Expression study for some key enzymes of metabolic pathways. RT-PCR products after exposure of cell cultures of *A. thaliana* to microgravity (TEXUS): transcripts as in Fig. 2.16

TEXUS 39 Microgravity Figure 2.18 shows expression profiles for the same genes given in Fig. 2.17 under short-term microgravity. In this case, a 1 *g* ground control was compared to a sample from the module before lift-off and at the end of the microgravity phase (about 6 min of μ*g*). The same transcripts as quantified under hyper-g did not exhibit any changes at all.

Clinorotation (3D) This trial of simulation of μ*g* was performed using a random positioning machine (RPM) at the ETH Zurich (group of Dr. Cogoli). As under TEXUS μ*g*, we could not detect any alterations in transcript levels between 15 min and 16 h of exposure (Fig. 2.18).

Magnetic Levitation Depending on the position within the magnet, 1 *g* and μ*g* could be achieved. We exposed A.t. callus cultures for up to 16 h in tubes at both positions. The respective transcription profiles are given in Fig. 2.19. As expression control the 18 S rRNA transcript was used. Compared to the regular 1 *g* control, no changes were found under both simulated μ*g* and 1 *g*.

2.3.3.2 Expression of Genes Coding for Transcription Factors

Transcription factors regulate gene expression and are manyfold regulated themselves. As we detected changes in the WRKY gene family of transcription factors in our array studies, we quantified the expression of some of these important transcripts

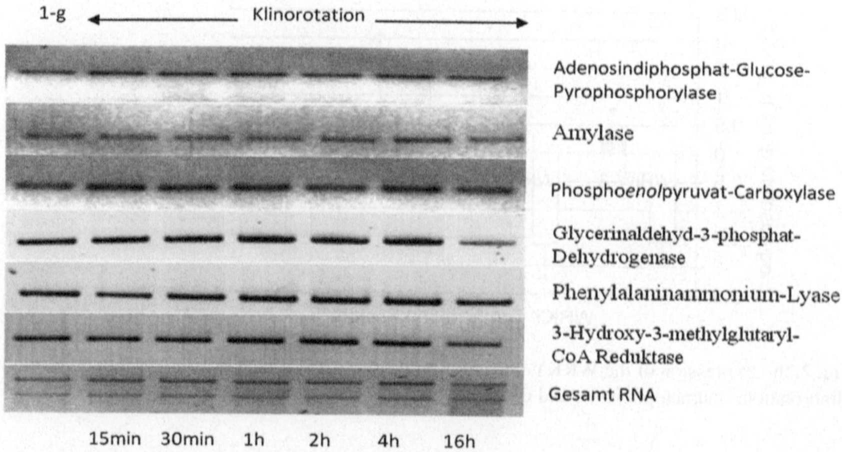

Fig. 2.18 Expression study for some key enzymes of metabolic pathways. RT-PCR products after exposure of cell cultures of *A. thaliana* to clinorotation. Transcripts as in Fig. 2.16

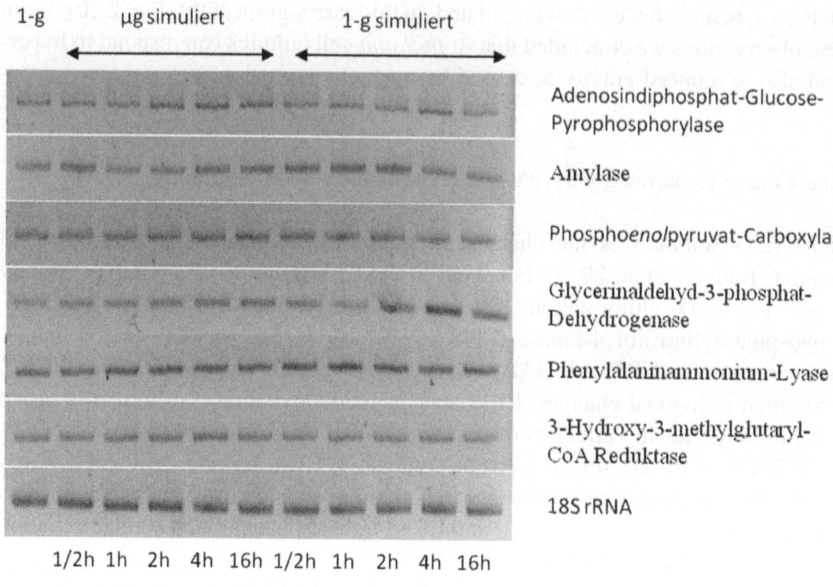

Fig. 2.19 Expression study for some key enzymes of metabolic pathways. RT-PCR products after exposure of cell cultures of *A. thaliana* to magnetic levitation. Transcripts as in Fig. 2.16

by real-time RT-PCR of differently exposed *Arabidopsis* cell cultures. The reproducibility was determined by the intra-assay variation with three repetitions within one PCR run. The inter-assay variation was given by three independent RNA extractions and three independent PCR runs. The former was less than 2.4%, the latter less than 4.5%.

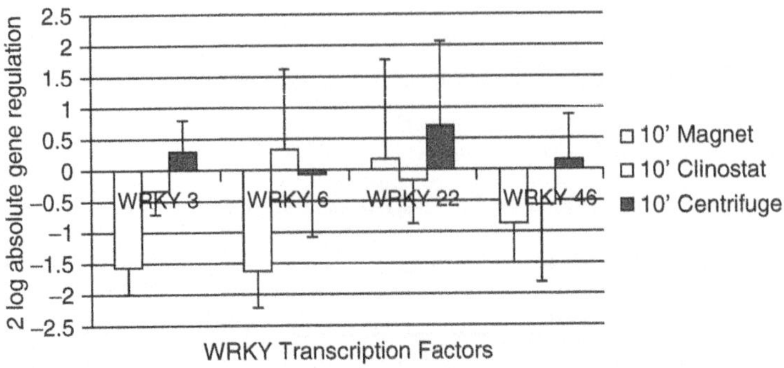

Fig. 2.20 Expression of the WRKY transcription factors 3, 6, 22, and 46 after 10 min exposure to clinorotation, magnetophoresis, and centrifugation (Babbick et al. 2005)

Ten minutes of clinorotation showed both downregulation (WRKY 22, 46, 6) and upregulation (WRKY 3, 70; Fig. 2.21). No alterations were significant. Ten min of hypergravity (8 *g*) increased the amount of transcripts coding for WRKY 22 and 46, although also not significant (Fig. 2.20). After 10 min of magnetic levitation, WRKYs 3, 6, and 46 were downregulated, in this case significantly (Fig. 2.20). From these observations we concluded that *A. thaliana* cell cultures can respond to hyper-*g* but also to reduced gravity as caused by magnetic levitation.

Time-Course Experiments: Hyper-*g* and Simulation of Microgravity

Figure 2.21 summarizes the changes in transcripts in relation to treatment and duration (Babbick et al. 2007). For 9 out of the 12 transcripts studied, hypergravity (8 *g*) and 2-D clinorotation resulted in nearly identical (AGL 84, ERF 5, phosphatidylinositol 4-kinase (PI4K), WRKYs 3, 22, 46, and 65) or mainly similar responses (ERF sub, IAA 19, and WRKY 70). Only MYB and WRKY 6 exhibited individual changes. RPM and magnetic levitation also formed a group with similar trends in variations of pool sizes for ERF 5, ERF sub, IAA 19, MYB, and WRKYs 3, 6, 22, 46, and 70. In contrast, WRKY 65 and AGL 84 differed considerably at times in their response to both treatments. However, there was also a general response in that WRKY 70 and ERF5 transcripts exhibited similar variation in amount for all four treatments.

The largest changes in expression resulted from the RPM and magnet treatments with between 2^3- and 2^5-fold increases in amounts of transcripts of ERF sub, IAA 19, WRKYs 3, 22, and 46. Changes upon exposure to centrifugation and clinorotation were less pronounced and ranged mostly between $+2^2$ and -2^1, MYB being an exception with about -2^5.

These groupwise similarities in response also became evident when the data were subjected to a correlation study (Table 2.2). Significant Pearson's correlations were shown between centrifugation and clinorotation ($p \leq 0.05$) and, to a greater extent, between magnetic levitation and random positioning ($p \leq 0.01$).

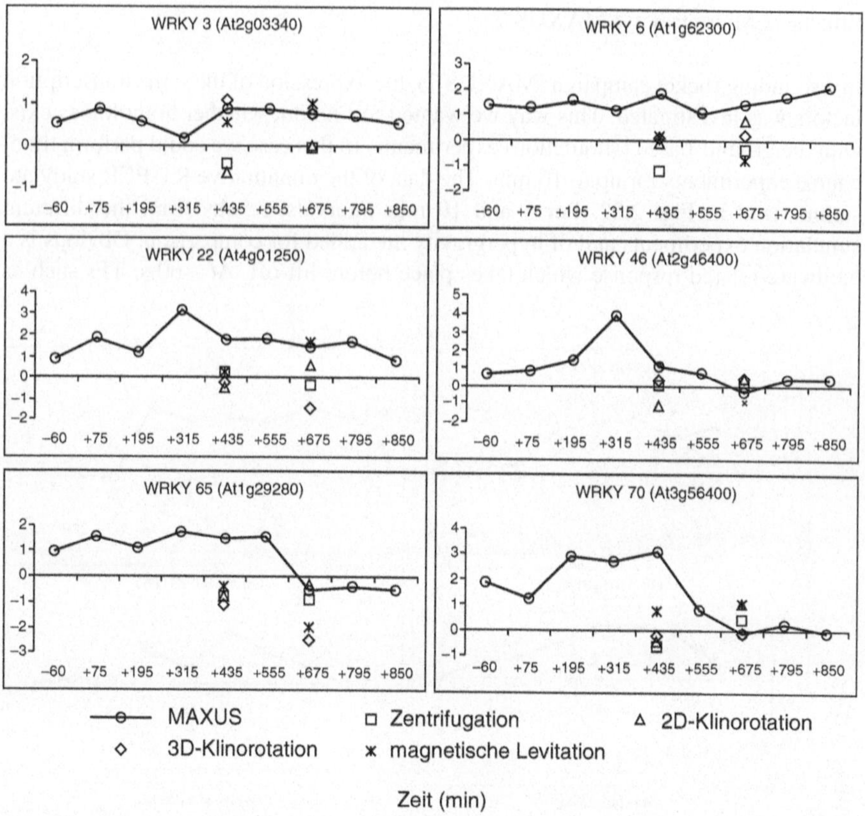

Fig. 2.21 Expression of WRKY transcription factors during a MAXUS sounding rocket experiment. Suitable time points are compared with responses to hypergravity (8 *g*) and simulated microgravity (2-D, 3-D clinorotation, magnetic levitation). Amounts of transcripts were quantified by real-time RT-PCR and calculated with REST®

Table 2.2 Groupwise comparison of expression of transcripts in *Arabidopsis* callus under different gravity conditions ($p < 0.05$)

	MAXUS	Hypergravity (8 *g*)	2-D clinostat	Random positioning machine	Magnetic levitation
MAXUS		0.084	−0.167	0.106	0.202
Hypergravity (8 *g*)	0.084		0.151	0.203	0.100
2-D clinostat	−0.167	0.151		−0.133	−0.417
Random positioning machine	0.106	0.0203	−0.133		0.207
Magnetic levitation	0.202	0.100	−0.417	0.207	

Effects of Microgravity: MAXUS 7

In a sounding rocket campaign (MAXUS 7), the expression of the same transcription factors was investigated. This way we wanted to find out whether homologies exist with the ground-based (simulation) experiments. In this case we could perform time-course experiments for up to 10 min. The data of the quantitative RT-PCR study are summarized in Fig. 2.22. For 5 and 10 min intervals, results from the different simulation experiments and of hypergravity are added for comparison. Obvious is a hardware-related response which takes place before lift-off. At −60 s, TFs such as

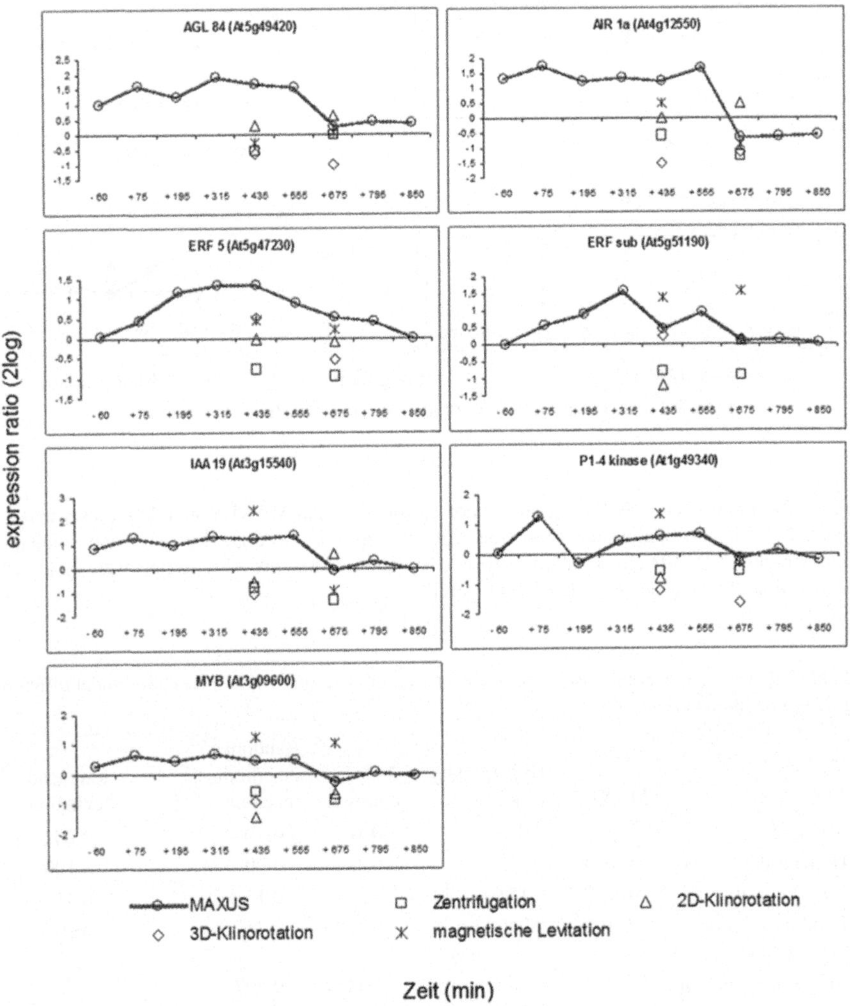

Fig. 2.22 Expression of non-WRKY transcription factors during a MAXUS sounding rocket experiment. For details compare Fig. 2.21

Table 2.3 Groupwise comparison of expression of transcripts in *Arabidopsis* callus under different gravity conditions

	Random positioning machine	Hypergravity (8 *g*)	2-D clinostat	Magnetic levitation
Random positioning machine				
Pearson's corr. coeff.	1	0.07	0.113	0.514**
Significance (two sided)		0.54	0.324	0.00
Hypergravity (8 *g*)				
Pearson's corr. coeff.	0.07	1	0.235*	0.33
Significance (two sided)	0.54		0.039	0.773
2-D clinostat				
Pearson's corr. coeff.	0.113	0.235*	1	0.144
Significance (two sided)	0.324	0.039		0.21
Magnetic levitation				
Pearson's corr. coeff.	0.514**	0.033	0.144	1
Significance (two sided)	0.00	0.773	0.21	

Similarities in responses were determined according to Pearson (SPSS software package). For this purpose all expression data shown in Fig. 2.22 were used
$n = 78$ throughout (Babbick et al. 2007)
*,**Correlations significant at a level of 0.05 or 0.01, respectively (two sided)

AGL 84, AIR 1a, and IAA 19 show a higher expression than the 1 *g* controls outside the module. This is also true for most of the WRKY TFs (Fig. 2.22). For a correlation study, the values for 5 and 10 min of exposure are compared (Table 2.3). These data yield highest correlations of MAXUS microgravity with magnetic levitation and random positioning.

The data indicated that (1) changes in gravity induce stress-related signaling, and (2) exposure in the RPM induced changes in gene expression, which resemble those of magnetic levitation. Two-dimensional clinorotation resulted in responses similar to those caused by hypergravity. It was thus suggested that RPM and ML are more suitable to simulate microgravity than clinorotation.

Functions of TFs

TFs are an important group of regulatory elements. They are small proteins which recognize DNA in a sequence-specific manner and which regulate the frequency of initiation of transcription upon binding to specific sites of the promoter of target genes (Stracke et al. 2001). TFs can be both activators and repressors depending on interactions with others and thus control or influence many biological processes, including responses to the environment. Many TFs autoregulate at the level of transcription (Chen et al. 2002). TFs have been classified into families based on sequence similarities, especially within DNA-binding domains. Major families of *Arabidopsis* transcription factors are MYB, AP2/EREBP, NAC, bHLH/MYC, bZIP, HB, MADS, WRKY, ARF-Aux/IAA, and Dof (Riechmann and Ratcliffe 2000), of which WRKY and ERF are plant-specific. Attention was thus focused on members of the latter two families, in addition to a MYB and a MADS-box family TF, and two members of signaling pathways, namely, IAA 19 and phosphoinositol 4-kinase (PI4K).

The WRKY family of TFs is involved in the control of stress responses due, for example, to infection or wounding (Eulgem et al. 2000). They are defined by a DNA-binding domain that contains the conserved amino acid sequence WRKY. After binding to their cognate W box binding motif, they activate transcription of the respective gene. Targets are regulatory genes such as receptor kinases. Of this family, expression was investigated for WRKYs 3, 6, 22, 46, 65, and 70. WRKY 70 is part of a salicylic acid (SA)-dependent signal pathway. Overexpression increases pathogen resistance (PR), but it can be both a positive and a negative regulator of PR gene expression. Thus, it is suggested that WRKY 70 is an important node of convergence integrating SA (activator) and jasmonic acid signaling (repressor; Li et al. 2004). In our investigation we could show that pool sizes increased slightly in all treatments. WRKY 6 has been classified as an immediate-early-type gene, which does not require de novo protein synthesis, and is involved in both regulation of plant defense and senescence. WRKY 6 can also function both as a positive and negative regulator of gene expression (Robatzek and Somssich 2001). This TF also showed a weak response to the different experimental treatments. WRKY 22 expression is induced after activation of a specific MAP kinase cascade (Asai et al. 2002), the latter being induced by diverse types of stressors, such as pathogens (Asai et al. 2002). The resulting overexpression of WRKY 22 leads to enhanced resistance to pathogens. WRKY 22, like WRKY 6, can stimulate its own expression. The expression of this gene was increased up to eightfold upon RPM and ML treatments, whereas there was only a slight response to hypergravity (8 *g*) and 2-D clinorotation.

Phosphatidylinositol 4-kinase (PI4K), to give another example, is part of a metabolic pathway, starting with phosphatidylinositol (PI), which is involved in plant responses to many environmental stimuli. More than 80 polypeptides have been identified as being involved in PI signaling (Lin et al. 2004). PI4K, with PI-phosphate 5 kinase, catalyzes the sequential phosphorylation of PI to $PI(4,5)P_2$, which is the substrate of phospholipase C. The latter is involved in Ca^{2+}-dependent signaling pathways (Perera et al. 2001; Lu and Feldman 1997). Many steps of the PI pathway are located in or near the plasma membrane and could be involved in the sensing of gravitational effects on the membrane/cytoskeleton complex. The PI4K gene family showed increased expression upon IAA, drought, and salt treatments (Lin et al. 2004). Expression was largely increased in the RPM, but there is also an increase (after an initial decline) by ML treatment. Recent microarray studies with *Arabidopsis* cell cultures exposed to microgravity for 5 min showed an increased expression of diacylglycerol kinase (Martzivanou et al. 2006). The latter synthesizes phosphatidic acid from diacylglycerol. Phosphatidic acid is an activator of protein kinase C. It is thus possible that random positioning, magnetic levitation, and microgravity induce similar Ca^{2+}-dependent signaling pathways.

2.3.3.3 Conclusion for Transcription Factor Studies

A number of identified TFs showed a steady increase in response to the experiment parameters assessed. However, only a few can be related to defined responses. This is due mainly to the multilevel interaction of TFs of the same or different families. The data indicates that a part of gravity-induced signaling cascades is also involved in other stress responses, mainly those related to plant-pathogen interactions. The challenge is to identify steps prior to these cascades, which are gravity-specific. One interesting finding of this study was that there was some degree of homology of response to the RPM and ML treatments, as well as those reported for hypergravity and 2-D clinorotation. This indicates that clinorotation could exert a continuous (although sequential) sensitizing of mechanoreceptors, which is not so different from centrifugation and which could explain the differences in comparison with a microgravitation environment (Link et al. 2001). The RPM, in contrast, by its random action is more likely to compensate 1 *g* effects in such a rapid way that cells "forget" such 1 *g* stimulation. For a comparison between microgravity and RPM, see Kraft et al. (2000).

2.3.4 Array Studies

2.3.4.1 Macroarrays ("Home Made")

Macroarrays, also called "reverse northern," were initially used for testing the expression pattern of diverse genes. For this purpose known cDNAs were fixed on a membrane and hybridyzed with poly-A^+-RNA from the respective sample (Gress et al. 1992). Comparison of the resulting signal intensities gives an idea about up- and downregulated gene transcripts. Genes of interest were selected according to a microarray study (Incyte Genomics, see below).

About 50 genes from different functional groups were investigated (Martzivanou et al. 2006). Poly-A^+-RNA was isolated from experiments dealing with hypergravity (8 *g*/40 and 60 min), clinorotation (2-D; 4 and 16 h), magnetic levitation (1 *g*/simulated 1 *g*; simulated µ*g* for 30 min and 1 h), and sounding rocket exposure (1 *g* control outside payload; 1 *g* control inside payload before takeoff; sample after 6 min of µ*g*). The hybridization signals were analyzed by a gel documentation system (BIO-RAD). The values for *GAPDH* and 18S rRNA were taken as controls, assuming that their transcripts were not affected by the different treatments. Differences were verified by RT-PCR.

Data from a sounding rocket experiment (TEXUS 39) showed that 6 min of microgravity had an effect on gene expression (Table 2.4; Fig. 2.23). Transcripts of the ammonium transporter AMT1 (link between nitrogen assimilation and carbon availability), arginine decarboxylase (AGC; response to environmental stress), pyruvate kinase (PK; carbon supply for amino acid synthesis), diacylglycerol kinase (DAGK, effect on protein phosphorylation), transcription factor WRKY 33 (see above), and of homoserine kinase (HSK; amino acid synthesis) were increased in amount. Similar analyses with samples from 2-D clinorotation and magnetic levitation showed no changes.

2.3.4.2 Early Microarrays (Incyte Genomics)

In order to get more information about *g*-related effects on gene expression, we used a 1 h exposure to 7 *g* for a microarray study, using a commercial *A. thaliana* chip with 4105 unique annotated clusters/genes (Incyte Genomics). At that time, this was the largest microarray available. With this array, we found transcripts of more than 200 genes to be significantly increased in amount (ratio 7 *g*/1 *g* control; $2^{1.6}$ and larger). Examples are given in Table 2.5. They fall into several categories.

Transcripts coding for enzymes of major pathways formed the largest group (25%), followed by gene products involved in cellular organization and cell wall formation/rearrangement (17%), signaling, phosphorylation/dephosphorylation (12%), proteolysis and transport (10% each), hormone synthesis plus related events (8%), defense (4%), stress response (2%), and gravisensing (2%). Many of the alterations were part of a general stress response, but some changes related to the synthesis/rearrangement of cell wall components could be more hyper-g-specific. We only found few gene products which were decreased in relation to 1 *g* controls, and these were less significant (ratio $<2^{1.6}$). We thus assumed that g-forces above a threshold of about 5 *g* for 1 h are sensed by plant cells in general, causing distinct metabolic responses which obviously, in part, are regulated by gene expression (Martzivanou and Hampp 2003).

2.3.4.3 Affymetrix Microarrays

With the advancement of the microarray technology, we were able to extend our expression analysis to more than 25,000 genes (Affymetrix GeneChip). We used this possibility to compare changes in transcript levels due to microgravity exposure for short (20 s, 6, 12 min) and long periods of time (5 days), as well as to partial g (0.16 and 0.38 *g*) or simulation of microgravity by clinorotation.

Table 2.4 Expression of a selected set of genes after short-term exposure of *A. thaliana* cell cultures to microgravity (TEXUS 39, 6 min), hypergravity (8 *g*), and clinorotation (three-dimensional)

		8 *g* (centrifugation)			Clinostat treatment			Microgravity		
Gene	Accession no.	Control	Hyper-*g*	Abundance (8 g/c)	Control	Clinostat	Abundance (clino/c)	Control	µg	Abundance (µg/c)
Ammonium transporter	X75879	4.7 ± 0.3	3.2 ± 0.7	0.69	4.6 ± 0.25	4.2 ± 0.7	0.9	1.0 ± 0.18	2.8 ± 0.13	2.9
Arginine decarboxylase	U52851	3.8 ± 0.25	3.6 ± 0.28	0.94	3.7 ± 0.05	1.9 ± 0.28	0.5	0.1 ± 0.01	0.9 ± 0.33	8.7
Diacylglycerol kinase	D63787	3.9 ± 0.36	3.4 ± 0.71	0.86	3.5 ± 0.66	5.3 ± 0.71	1.5	0.2 ± 0.12	1.5 ± 0.13	6.4
Homoserine kinase	AF082525	3.5 ± 0.51	3.6 ± 0.33	1.02	4.3 ± 0.24	5.0 ± 0.33	1.2	1.1 ± 0.04	2.2 ± 0.29	2.1
MAP kinase	AAC67338	nd			nd			0.6 ± 0.34	1.9 ± 0.05	3.1
Phosphoinositol-spec. phospholipase C	D38544	4.0 ± 0.44	3.8 ± 0.08	0.93	3.6 ± 0.88	3.5 ± 0.08	1.0	4.8 ± 0.4	13.1 ± 0.63	2.7
Pyruvate kinase	BAB03043	7.1 ± 0.56	5.8 ± 0.21	0.81	7.1 ± 0.84	9.6 ± 0.21	1.4	0.3 ± 0.46	1.7 ± 0.39	5.3
Receptor-like kinase	AF001168	4.1 ± 0.21	2.5 ± 1.58	0.6	4.1 ± 0.58	4.4 ± 1.58	1.1	2.1 ± 1.66	5.1 ± 1.23	2.4
WRKY 33	AF509499.1	3.6 ± 0.14	2.6 ± 0.35	0.72	4.4 ± 0.8	4.4 ± 0.35	1.0	0.2 ± 0.01	2.4 ± 0.27	9.4

Abundance denotes the ratios of spot intensities of the respective treatment/1 *g* control

Data represent means of spot intensities (% adjusted volume, PD-QUEST software, BIO-RAD) ± SD (Martzivanou et al. 2006)

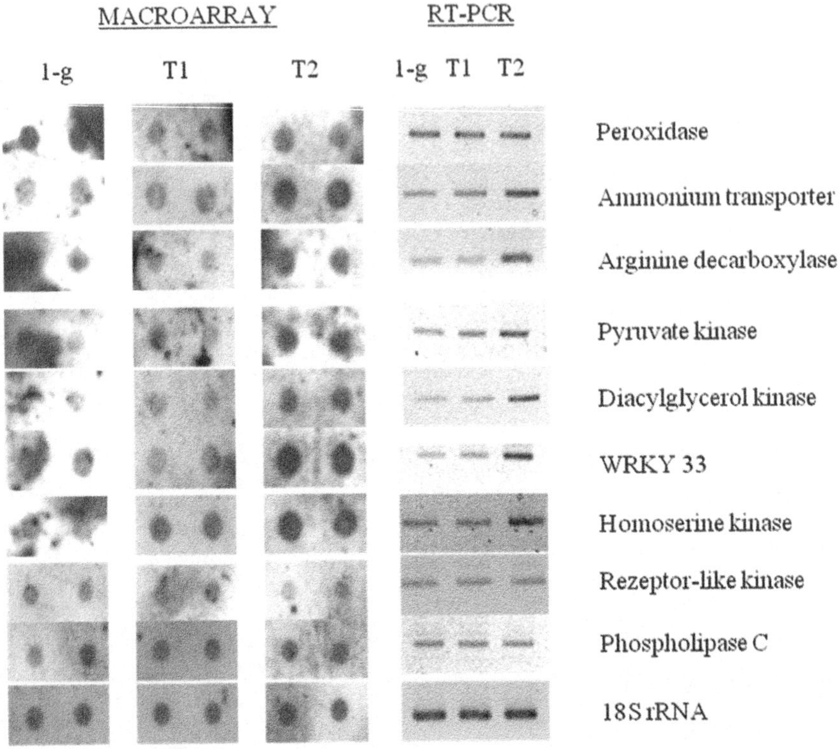

Fig. 2.23 Quantitation of selected transcripts by macroarrays and RT-PCR. *A. thaliana* cell cultures from a sounding rocket flight (TEXUS). *1-g* ground control; *T1* flight sample 10 min before lift-off; *T2* flight sample after 6 min of microgravity. *GAPDH* glyceraldehyde 3-phosphate dehydrogenase; *AMT1* ammonium transporter; *AGC* arginine decarboxylase; *PK* pyruvate kinase; *DAGK* diacylglycerol kinase; *WRKY* WRKY 33; *HSK* homoserine kinase; receptor-like kinase; *PLC* phospholipase C. As reference 18S rRNA was used (Martzivanou 2004)

2.3.5 Platforms

2.3.5.1 Short-Time μ*g* (Parabolic Flights)

Wild-type *A. thaliana* cell cultures were manually fixed by the injection of RNAlater (Ambion) at the end of every gravitational phase (1 *g*, 1.8 *g*, μ*g*, 1.8 *g*). In order to avoid an interference in responses by preceding variations in gravity, we specifically investigated the first parabola of each flight day. A microarray (ATH1/Affymetrix) screening resulted in specific gene expression profiles due to the different *g*-forces, although the largest part of the up- (more than twofold) or downregulated genes (less than 50%) responded to each force (Hausmann et al. 2014).

These transcripts appear to be unspecific to gravity-dependent processes and belong mainly to general stress responsive genes. Interestingly, we found also genes, which were specifically expressed after a given *g*-force. For example,

Table 2.5 Selection of transcripts/genes with increased expression after 7 *g* for 1 h, as compared with a 1 *g* control

Induced genes		Ratio
Gene activation/transcription	Histone acetyltransferase	2.3
	Initiation factor 4B	1.7
	Transcription factor	1.7
	Protein synthesis factor	1.7
	Scarecrow	1.7
	RNA polymerase II, subunit	1.6
Cellular organization/biogenesis	Cytochrome P450	1.8
	UDP-glucose glycosyltransferase	1.8
	Syntaxin	1.7
	Kinesin	1.7
	Actin	1.7
Cell wall biosynthesis	Arabinogalactan protein reorganization	2.2
	Fructose bisphosphatase	2.0
	Extensin	1.9
	Xylosidase	1.7
	Cellulose synthase	1.6
Signaling	Diacylglycerol kinase	1.7
	MAP-kinase	1.7
	Phospholipase D	1.7
	Calcium binding protein	1.6
	Phosphatidylinositol synthase	1.6
Protein kinases/phosphatases	Serine/threonine protein kinase	1.8
	SHAGGY protein kinase	1.6
	Receptor protein kinase	1.6
	Protein phosphatase 2C	1.6
	ABI2	1.6
Proteases	Subtilisin similar	2.1
	Ubiquitin-specific protease	1.8
	Thiol protease	1.7
	Carboxypeptidase	1.7
	Cysteine proteinase	1.7
Hormones	ACC oxidase	1.9
	Methionine synthase	1.9
	HOOKLESS 1	1.7
	Protein phosphatase 2A	1.7
	Transcription activator	1.9
Stress	Catalase	2.0
	Cytochrome P450	1.8
	Glutamate decarboxylase	1.8
	SRG 1	1.7
	HSP90	1.7

(continued)

Table 2.5 (continued)

Induced genes		Ratio
Defense	Cinnamoyl-CoA reductase	1.8
	NPR 1	1.8
	1,3-glucanase	1.8
	ELI3	1.6
	Anthranilate benzoyl transferase	1.7
Gravity	Peroxidase	1.9
	Ca-ATPase	2.0
	Protein phosphatase 2A	1.8
	ARG 1	1.7
	Protein phosphatase I	1.6
Transporter	Phosphate transporter	2.4
	Ammonium transporter	2.1
	H^+-ATPase	1.9
	Amino acid transporter	1.8
	Adenine nucleotide translocase	1.7
Primary metabolism	Aconitase	1.8
	Pyruvate kinase	1.8
	Malate oxidoreductase	1.7
	Malate dehydrogenase	1.7
	Fructose bisphosphate aldolase	1.6
Amino acid metabolism	Arginine decarboxylase	2.2
	Glycine dehydrogenase	2.0
	Homoserine kinase	1.7
	Glutamine synthetase	1.7
	Phosphoserine aminotransferase	1.7
Secondary metabolism	Choline monooxygenase	1.7
	Ferrochelatase	1.7
	Formate dehydrogenase	1.7
	AMP-deaminase	1.6
	Lipoxygenase	1.6

Ratios of more than 1.6 indicate a significantly different expression

396 transcripts were at least 100% upregulated during the microgravity phase of the parabola, among them 25 Ca^{2+}-dependent genes, such as members of the Ca^{2+}-binding protein family, or Ca^{2+}-dependent protein kinases (Table 2.6).

2.3.5.2 Partial *g* (Parabolic Flights and Simulation by Clinorotation)

With regard to a possible future habitation of Mars or Moon, parabolic flights with partial-*g* forces were offered by the German Aerospace Center (DLR, Germany) together with CNRS (France) and Novespace (Bordeaux, France) (Joint European Partial-*g* Parabolic Flight, JEPPF, 0.38 *g*, 0.16 *g*). For comparison, we also simulated

Table 2.6 Selection of transcripts increased in amount by more than 100% after 20 s of microgravity (signaling, Ca^{2+}-related processes, detoxification of ROS)

Gene identifier	Gene description	Protein function
AT4G34150	Calcium-dependent lipid-binding (CalB domain) family protein	Calcium-dependent membrane targeting
AT1G02270	Calcium-binding family	Calcium ion binding
AT1G05630	Inositol polyphosphate 5-phosphatase	Ins(1,4,5)P3-dependent calcium release
AT1G35670	Calcium-dependent protein kinase 2 (CDPK2)	Calcium sensor: MAPK-cross talk
AT1G20630	Catalase 1 (CAT1)	Detoxification of hydrogen peroxide
AT2G28190	Copper/zinc superoxide dismutase 2 (CSD2)	Detoxification of superoxide radicals
AT1G14920	Transcription factor	Reduces ROS accumulation by upregulating the transcription of superoxide dismutases
AT2G43350	Glutathione peroxidase 3 (GPX3)	Functions as both a redox transducer and a scavenger in stress responses
AT4G35290	Glutamate receptor 2 (GLUR2)	Cellular calcium ion homeostasis
AT5G11210	Ligand-gated ion channel	Cellular calcium ion homeostasis

partial *g* in clinostat experiments (fast-rotating clinostat, compare Shen-Miller et al. 1968; Sobick and Sievers 1978; 0.38 *g*, 0.16 *g*, and 0.5 *g* were tested). These experiments were also part of an attempt to identify the potential threshold acceleration for gravity-dependent molecular responses.

Figure 2.24 (Fengler et al. 2015a) shows that cell cultures respond to 0.16 *g*, 0.38 *g*, and 0.5 *g* during parabolic flights and ground-based experiments to different extents.

Interestingly, under partial *g* the number of at least twofold differentially expressed genes (DEGs) was much lower than under short-term μ*g* during a regular parabolic flight (thousands of DEGs during the first parabola (Neef et al. 2011, 2013b; Hausmann et al. 2014)). The sequence of parabolas was such that they started with 0.38 *g*, followed by 0.16 *g*, and ended with μ*g*. The slightly higher response to 0.38 *g* in comparison with 0.16 *g* could thus be due to this sequence. At the first exposure to 0.16 *g*, the cells had already experienced 12 0.38 *g* parabolas.

In the clinostat experiment (Fig. 2.24), there was no clear difference between the response to 0.16 *g* and 0.38 *g*, with no obvious time-dependent changes over 30 min. In contrast, simulated 0.5 *g* (not available under flight conditions) resulted in a clear decrease in DEGs from 5 to 30 min. A simulation of the parabolic flight as shown under C^2 (33 s; with or without preceding 1.8 *g*; Fig. 2.24) resulted in a considerably higher number of altered transcripts. This finding suggests a decreasing number of altered transcripts with increasing time of exposure.

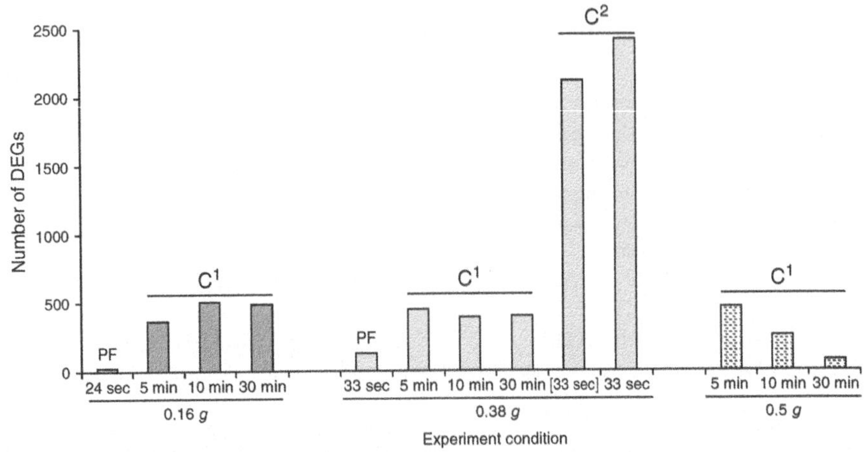

Fig. 2.24 Number of differentially expressed genes of *A. thaliana* cell cultures after exposure to partial-g parabolic flights and 2-D clinorotation. PF, parabolic flight; C, clinostat experiments; C①, clinorotation only; C②, clinorotation with preceeding 1.8 *g* treatment for 33 s (Fengler et al. 2015a)

2.3.5.3 Effect of Simulated Partial-*g* Conditions on Gene Expression

The calculation of the Pearson's correlation coefficient (CorP) showed that the effect of the different experimental conditions on the whole transcriptome was relatively low (response of a small fraction of genes). Considering the total number of generated gene expression values (about 23,000), this resulted in a high CorP (Fengler et al. 2015a). The clustering of expression values (again based on 23,000 genes) suggests that in the clinorotation experiment, gene expression was more affected by the time of exposure (clusters I–III, compare Fig. 2.24), than by the change in gravity (0.16 *g*, 0.38 *g*, or 0.5 *g* generate no separate cluster for a given time of exposure) (Fig. 2.25). The parabolic flight data form their own cluster (cluster IV). Interestingly, under clinorotation, data from 30 min of exposure clusters with that of 1 *g* controls. This indicates that effects of clinorotation probably start to disappear after that time.

The experiment from Fig. 2.24 was focused on the simulation of 0.38 *g*. To be as close to parabolic flight conditions as possible, we tested both the effect of simulated 0.38 *g* for 33 s (clinorotation only) and the same treatment with a preceding 1.8 *g* phase which lasted for 22 s ([C^2]). The response within the seconds range was more prominent than that upon prolonged time of exposure (Fig. 2.24). Nearly 2500 genes (0.38 *g* simulation only) were differentially expressed (Fig. 2.24). The preceding hyper-*g* treatment ([C^2]) reduced the number of DEGs (2110 DEGs). In detail, 2414 DEGs were differentially expressed under C^2 conditions [1267 upregulated and 1147 downregulated (Fengler et al. 2015a)]. Less DEGs (2110) were altered with preceding hyper-*g* treatment ([C^2]) (1102 up- and 1008 downregulated). The effects induced by both short-time simulation experiments were, however, very similar (C^2 and [C^2]). 1.872 DEGs were identical (995 upregulated, 877 downregulated). A

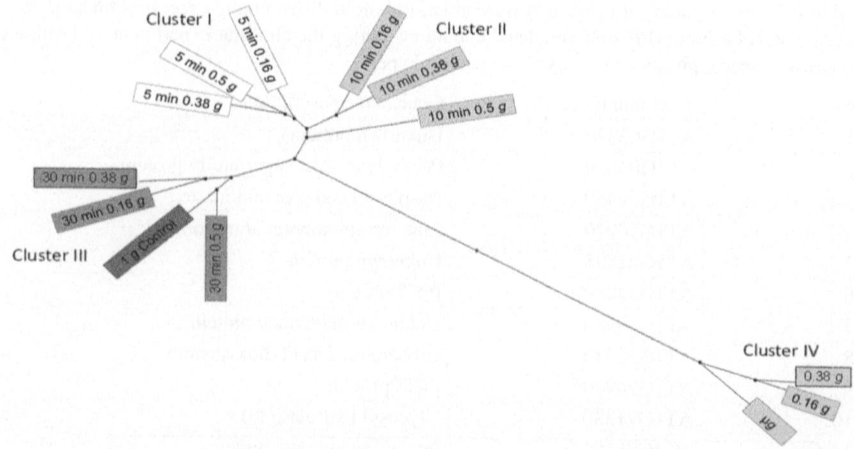

Fig. 2.25 Unrooted hierarchical clustering of gene expression of *A. thaliana* cell cultures exposed to real (parabolic flight; cluster IV) and simulated (clinostat, C①, clusters I–III) partial *g* (Fengler et al. 2015a)

minor group of altered genes were specific for either treatment (C^2 542 DEGs, ([C^2]) 238 DEGs). With regard to the effect of the simulation, we were especially interested in genes affected by short-time clinorotation. These also comprise candidates which were identical under "real" and simulated conditions (44 and 68 DEGs; Fengler et al. 2015a).

2.3.5.4 Simulated and "Real" Partial-*g* Forces: A Comparison

The clustering in Fig. 2.25 illustrates the low similarity between "real" (as experienced during PF) and simulated partial *g* within the range of minutes. This is shown by the separation of clusters generated from parabolic flights (cluster IV) and clinostat samples (clusters I–III). In addition, the correlation showed a time-dependent decrease with the weakest correlation after 30 min. However, screening at the single gene level identified many similarities (i.e., identical DEGs) between "real" and simulated conditions. This was mainly the case for the seconds range of exposure to 0.38 *g*. Out of 127 DEGs which responded during flight conditions (Fig. 2.24), 68 genes could also be induced under simulated conditions (54%, C^1). In case of the preceding hyper-*g* treatment, this number was reduced (44 identical genes). Among these genes (68 and 44) 43 were identical and are listed in Table 2.7. Among them are gene candidates that code for proteins that are known to be involved in the modification of cell wall properties like pectate lyase-like membrane proteins, glycosyl hydrolases, a xyloglucan endotransglucosylase, and proteins that are not characterized yet. Additionally, there are peroxidases, auxin-related proteins (e.g., PINs), transmembrane proteins, and components of intracellular signaling chains like different protein kinases, transcription factors, and phytohormone-related proteins.

Table 2.7 List of identical genes that were at least twofold differentially expressed under 0.38 *g* during parabolic flight (PF) and simulated conditions during the clinostat experiment C^1 (without preceding hyper-*g* phase) and C^2 (with short-term hyper-*g*)

No	ATG number	Gene name/description
1	AT1G03820	Unknown protein
2	AT1G04680	Pectin lyase-like superfamily protein
3	AT1G09450	Haspin-related protein kinase
4	AT1G10970	Zinc transmembrane transporter 4
5	AT1G22885	Unknown protein
6	AT1G23080	PIN7 protein
7	AT1G29980	Unknown membrane protein
8	AT1G30040	Gibberellin 2-beta-dioxygenase
9	AT1G70940	PIN3 protein
10	AT1G71380	Glycosyl hydrolase 9B3
11	AT1G75500	Integral membrane protein
12	AT2G07751	NADH:ubiquinone/plastoquinone oxidoreductase
13	AT2G15830	Unknown protein
14	AT2G32990	Glycosyl hydrolase 9B8
15	AT2G40480	Unknown protein
16	AT2G43520	Trypsin inhibitor protein 2
17	AT2G47860	Phototropic-responsive protein
18	AT3G08030	Unknown protein
19	AT3G14205	Phosphoinositide phosphatase protein
20	AT3G24503	Aldehyde dehydrogenase
21	AT3G54260	Trichome birefringence-like protein
22	AT3G56170	Ca^{2+}-dependent nuclease (CAN)
23	AT4G03210	Xyloglucan endotransglucosylase (XTH) 9
24	AT4G14630	Germin-like protein 9
25	AT4G15630	CASP-like protein 1E1
26	AT4G21960	Peroxidase 1
27	AT4G24780	Pectate lyase protein
28	AT4G25200	Heat shock protein 23.6
29	AT4G36110	Small auxin-related protein
30	AT4G39320	Microtubule-associated protein
31	AT5G05340	Peroxidase 52
32	AT5G12340	Unknown protein
33	AT5G13205	Copia-like retrotransposon family
34	AT5G16250	Unknown protein
35	AT5G22270	Salt-induced/ein3/eil1-dependent protein
36	AT5G26220	ChaC-like protein
37	AT5G41810	Unknown protein
38	AT5G46050	Peptide transporter protein 3
39	AT5G46590	NAC domain-containing transcription factor
40	AT5G61480	Leucine-rich repeat transmembrane kinase
41	ATCG00020	Photosystem II reaction center protein a

(continued)

Table 2.7 (continued)

No	ATG number	Gene name/description
42	ATCG00420	NADH dehydrogenase subunit protein
43	ATCG00430	NADH dehydrogenase subunit protein

Genes are represented by their gene identifier and gene name/description (Fengler et al. 2015a)

2.3.5.5 Evaluation of Partial-*g* Data

The identified gene expression pattern is considered to be typical for the initial steps in sensing partial-*g* forces. In contrast to earlier studies, where only parabolas yielding microgravity were flown (over 5000 DEGs), the number of DEGs under parabolic flight partial *g* was low during the first parabola (compare Hausmann et al. 2014). In the case of regular μg flights, we always only analyzed the first parabola of a given flight day, where the cells experienced microgravity for the first time. In the partial-*g* study, this was different. As the partial-*g* parabolas were flown in a sequence which started with 0.38 *g*, only the cells of this group experienced reduced gravity for the first time. Those exposed to the first 0.16 *g* parabola had experienced 13 0.38 *g* parabolas before, those with μg (not shown), additional 13 0.16 *g* parabolas. This could explain the decreasing response.

We found a considerable range of regulated gene products that were identical under PF and simulated conditions (Table 2.7). One important overlap is an impact on cell wall organization. The majority of identical candidates (PF and C[1]) are involved in the modification of cell wall components, like pectate lyase-like membrane proteins (AT1G04680, AT4G24780; upregulated). This probably decreases the mechanical strength of the primary cell wall by cleaving the α-1,4-linked galacturonic acids which leads to a depolymerization of pectins (Sun and van Nocker 2010). Also upregulated were glycosyl hydrolases 9B3 (also named cellulase 3, AT1G71380) and 9B8 (AT2G32990), as well as a xyloglucan endotransglucosylase 9 (AT4G03210). These proteins are located in the plasma membrane or in the cell wall and are able to cleave the glycosidic bonds of polysaccharides. This can lead to cell wall expansion by polysaccharide degradation (Minic and Jouanin 2006).

As shown for the simulation of reduced *g*-forces by magnetic levitation (Manzano et al. 2012), the expression of peroxidase genes [peroxidase 52 (AT5G05340), peroxidase 42 (AT4G21960)] is also upregulated under such conditions. Peroxidases are involved in the detoxification of reactive oxygen species in the extracellular matrix, which always are part of stress responses. PIN3 (AT4G36110) and PIN7 (AT1G23080), two auxin-related proteins that are known to be sensitive to gravitational changes (Aubry-Hivet et al. 2014), as well as the small auxin-responsive protein (AT4G36110), are differentially expressed under both "real" and simulated 0.38 *g*. Furthermore, diverse integral membrane components appeared to be affected by both treatments, like the proteins of AT4G15630 and AT1G75500 and different transmembrane peptide transporters (AT1G10970, AT5G46050). In addition, members of intracellular signaling pathways appeared to be sensitive to partial *g* and changed in transcript abundance upon both treatments. These are, e.g., the NAC domain-containing transcription regulator (ANAC096, AT5G46590) located in the nucleus, the haspin-related protein kinase (AT1G09450), an inositol phosphatase (AT3G14205), and a transmembrane protein kinase (AT5G61480). This leucine-rich repeat kinase is involved in phytohormone-related signaling pathways, like the cytoplasm-located gibberellin 2-beta-dioxygenase (AT1G30040). Finally, the altered expression of reaction center proteins of the photosystems show that components of photosynthetic (ATCG00020, ATCG00420, ATCG00430) and the respiratory electron transport (NADH ubiquinone/plastoquinone oxidoreductase, AT2G07751) are affected by partial gravity levels as well. Processes induced especially under clinorotation were transport, development, protein metabolism, signal transduction, and transcription. In addition, we found a considerable number of genes involved in the posttranslational modification of proteins, such as diverse leucine-rich repeat membrane-bound kinases.

Owing to the sequence of parabolas, the partial-*g* parabolic flights can obviously not be used for the investigation of gravitational thresholds in *Arabidopsis*, due to the short and repetitive time of exposure to altered accelerations. Simulation by clinorotation of different gravitational field strengths revealed the least effects after 30 min at 0.5 *g*. After this period, the number of DEGs was lowest. We thus assume that after this time of clinorotation, specific responses are fading.

The 68 DEGs which were detected to be identical between real and simulated conditions are out of 23,000 genes (0.6%). This is not much but obviously specific for 0.38 *g*, because it is different from that induced upon μ*g* in similar cell culture material (Hausmann et al. 2014; Fengler et al. 2015a).

Taken together we conclude that molecular responses to partial *g* are in part different from those induced by μ*g* and that clinorotation can induce some changes in levels of transcripts similar to flight experiments.

2.3.5.6 Long-Time μg

This experiment was part of the SIMBOX (*S*cience *in* *M*icrogravity *Box*) mission, a joint project between the space agencies of Germany (Deutsches Zentrum für Luft- und Raumfahrt, Raumfahrtmanagement) and China (China Manned Space Engineering Office, now CMSA China Manned Space Agency) in November 2011. As one out of 17 biological experiments, callus cultures of *Arabidopsis* were exposed to a 17-day spaceflight mission in the SIMBOX incubator on board the Chinese spacecraft Shenzhou 8. Due to reduced vitality that had to be expected after longer periods of exposure within the flight hardware, the callus cultures were metabolically quenched after 5 days in space (Fengler et al. 2015b).

Owing to the availability of an in-flight centrifuge, it was possible to compare expression data with (a) real microgravity samples (thought to yield the microgravity-related alterations) and (b) with those from the ground controls (which should deliver effects of non-microgravity-related spaceflight conditions). This was achieved with high-density oligonucleotide arrays.

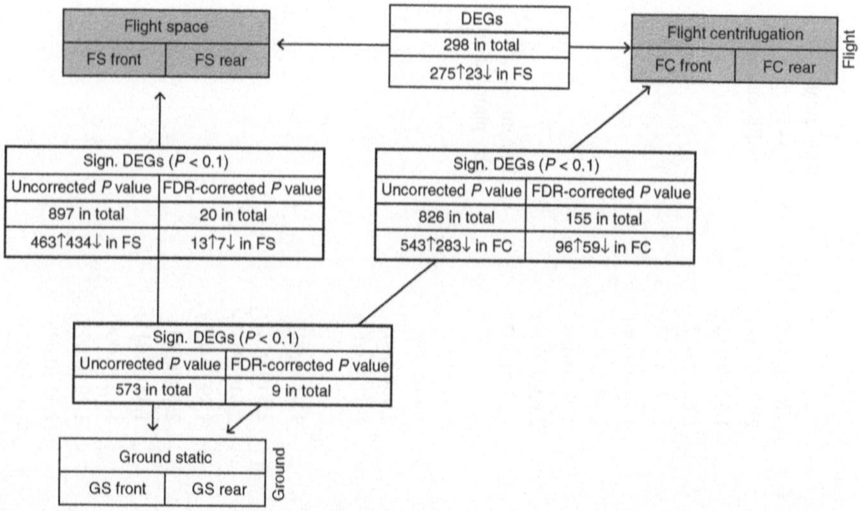

Fig. 2.26 Differentially expressed genes (DEGs, $P < 0.1$) between the flight and ground experiments with *A. thaliana* cell cultures exposed to 5 days of microgravity on board of Shenzhou 8. The different sample groups are identified by boxes. Arrows in boxes indicate numbers of up- or downregulated transcripts (Fengler et al. 2015b)

For detection of gene expression changes due to μ*g* exposure, we compared data generated from the sample groups flight space (FS) and in-flight centrifugation (FC). 275 genes were at least twofold differentially upregulated and 23 downregulated (Fig. 2.26). The application of statistics showed that there were no significant ($p < 0.1$) alterations at the expression level after 5 days in space. By means of a Gene Ontology-based Gene Set Enrichment Analysis (GSEA), the DEGs were related to common biological processes (Ashburner et al. 2000).

In order to identify processes which are specifically influenced by microgravity conditions, we compared overrepresented processes that were identical between sample group FS versus FC and FS compared to GS (Fengler et al. 2015b). Most prominent were effects on the translation machinery. Interestingly, all genes that were differentially upregulated and involved in translation processes were chloroplast-encoded. The other part of identified differentially upregulated genes is involved in electron transport chains located within mitochondria. Within the 23 downregulated genes (at least twofold), no special gene sets could be found, but the largest group codes for heat shock proteins.

2.3.5.7 Attempt to Distinguish Between Effects of Microgravity- and Non-microgravity-Related Spaceflight Conditions on Gene Expression

Another aim of this investigation was to separate responses to microgravity from those of non-microgravity-related spaceflight conditions. Only marginal data exist

Table 2.8 Differentially (at least twofold) and significantly expressed genes ($p < 0.1$, 573 in total) that are identical between flight space (FS) as well as in-flight centrifugation (FC) compared to ground static (GS)

No	ATG number	Gene name/description	log(fc) (p-value) FS vs. GS	log(fc) (p-value) FC vs. GS	Biological process
1	AT1G01560	MAP kinase 11	1.83 (0.034)	2.13 (0.027)	MAPK cascade
2	AT1G73500	MAP kinase 9	1.34 (0.006)	1.37 (0.038)	
3	AT3G01830	Calcium-binding EF-hand family protein	1.3 (0.032)	1.85 (0.027)	
4	AT3G47480		1.24 (0.081)	1.87 (0.037)	
5	AT2G40750	WRKY 54	1.29 (0.008)	1.74 (0.006)	
6	AT3G56400	WRKY 70	1.85 (0.008)	2.19 (0.004)	
7	AT5G22570	WRKY 38	2.52 (0 006)	3.28 (0.004)	
8	AT3G15500	NAC domain-containing transcription factor 3	2.98 (5.72E-4)	2.63 (0.003)	
9	AT1G35670	Calcium-dependent calmodulin-independent protein kinase 2	1.2 (0.002)	1.23 (0.003)	Protein phosphorylation
10	AT1G20650	Serine/threonine protein kinase family protein	−1.4 (0.024)	−1.5 (0.018)	
11	AT3G61160		−1.22 (0.007)	−1.4 (0.008)	
12	AT1G78290		1.71 (0.019)	2.0 (0.034)	
13	AT4G18640		1.08 (0.019) 1.07(0 014)	1.07 (0.014)	
14	AT4G18950		2.53 (0.031)	3.17 (0.02)	
15	AT5G16900	Leucine-rich repeat protein kinase family protein	1.42 (0.023)	2.0 (0.012)	
16	AT1G51890		2.55 (0.05)	2.64 (0.047)	
17	AT4G11480	Cysteine-rich receptor-like protein kinase family protein	1.56 (0.05)	1.89 (0.033)	
18	AT4G23260		1.65 (0.068)	2.49(0.041)	
19	AT4G38470	Tyrosine kinase family protein 46	1.14 (0.008)	1.34 (0.015)	
20	AT1G69790	Protein kinase superfamily protein	1.19 (0.038)	1.12 (0.009)	
21	AT5G53450	Protein kinase	1.88 (0.088)	1.89 (0.075)	
22	AT1G51620	Protein kinase family protein	1.8 (0.052)	2.31 (0.048)	
23	AT3G04530	Phosphoenolpyruvate carboxylase kinase 2	−1.6 (0.06)	−1.19 (0.43)	
24	AT5G63650	Protein kinase 2.5	−1.26 (0.028)	−1.01 (0.032)	

25	AT1G16260	Cell wall-associated protein kinase family protein	1.73 (0.006)	2.13 (0.003)	
26	AT1G68690	Proline-rich extensin-like receptor kinase family protein	1.04 (0.002)	1.03 (0.04)	
27	AT5G46330	Flagellin 2-induced receptor-like kinase	−1.85 (0.043)	−2.38 (0.016)	Defense response
28	AT2G19190		2.48 (0.065)	2.27 (0.075)	
29	AT2G15120	Disease resistance protein	2.68 (0.035)	2.53 (0.04)	
30	AT1G59780		1.37 (0.092)	1.98 (0.052)	
31	AT1G63880		−1.81 (0.003)	−1.79 (0.016)	
32	AT2G39200	Transmembrane domain-containing protein, similar to mildew resistance protein 12	2.6 (0.059)	2.55 (0.063)	
33	AT1G19610	Pathogenesis-related protein 14	−2.17 (0.002)	−2.14 (0.029)	
34	AT3G20600	Non-race-specific disease resistance protein	1.05 (0.037)	2.12 (0.011)	
35	AT1G02360	Chitinase family protein	2.6 (0.026)	2.87 (0.019)	
36	AT3G54420	Chitinase family protein class IV	1.73 (0.055)	2.53 (0.026)	
37	AT4G21390	Serine/threonine protein kinase family protein	1.5 (0.068)	1.86 (0.031)	
38	AT3G46280	Protein kinase family protein	1.83 (0.074)	2.3 (0.048)	
39	AT5G35750	Histidine kinase 2	−1.21 (0.042)	−1.35 (0.026)	
40	AT2G37130	Peroxidase 21	−3.06 (0.014)	−3.4 (0.008)	Response to oxidative stress
41	AT1G14540	Peroxidase 4	3.35 (0.018)	3.28 (0.02)	
42	AT5G05340	Peroxidase 52	2.15 (0.014)	2.06 (0.019)	
43	AT4G37530	Peroxidase family protein	2.17 (0.035)	2.15 (0.026)	
44	AT2G41480	Peroxidase 25	−1.04 (0.011)	−1.06 (0.034)	
45	AT1G20620	Catalase 3	−1.12 (0.07)	−1.39 (0.052)	
46	AT2G29490	Glutathione S-transferase 19 class tau 1	1.75 (0.07)	1.7 (0.073)	
47	AT3G22370	Oxidase family protein	1.3 (0.013)	1.0 (0.089)	
48	AT4G37220	Stress-responsive protein	2.87 (0.004)	1.87 (0.049)	Response to stress
49	AT4G21870	Heat shock protein 26.5	−1.31 (0.002)	−1.43 (0.012)	
50	AT2G38750	Calcium-dependent phospholipid-binding protein	1.48 (0.02)	1.15 (0.032)	

Changes are due to non-microgravity-related spaceflight conditions. The genes are sorted according to the overrepresented biological processes identified by GSEA to be most prominent (Fengler et al. 2015a)

about these effects on plants in space. Thus, we screened for genes that were significantly ($p < 0.1$) altered within spaceflight samples (FS and FC) compared to the 1 *g* ground control (GS) and identical between FS and FC compared to GS (Fengler et al. 2015b). This overlap yielded 573 significantly altered ($p < 0.1$) DEGs (Fig. 2.26). The majority of these genes could be related to intracellular signaling pathways such as mitogen-activated protein kinase (MAPK) cascades and protein phosphorylation. Included were different MAP kinases (e.g., AT1G01560, AT1G73500), serine/threonine/tyrosine kinases (e.g., AT1G20650, AT5G16900, AT4G38470), as well as many other kinases (Table 2.8). Furthermore, we identified genes coding for members of the calcium-binding EF-hand protein family (AT3G01830, AT3G47480) and the WRKY transcription factors 54, 70, and 38 (AT2G40750, AT3G56400, AT5G22570) that have also transcription regulation activity. Additionally, the spaceflight environment other than microgravity had a significant ($p < 0.1$) impact on general stress-responsive and defense-related genes, especially those involved in the response to oxidative stress and respiratory burst responses. Furthermore, general metabolic processes (gene set number 7), protein targeting, and rRNA processing were overrepresented due to non-microgravity-related conditions in space.

2.3.5.8 Evaluation of Long-Time μ*g* Data

The expression data of *Arabidopsis* callus cultures shows alterations in differential gene expression in response to microgravity. However, the influence of the spaceflight environment, in addition to microgravity, is significant.

Comparison between microgravity and 1 *g* space controls revealed about 298 differentially (but not significantly) expressed genes. This number is low in comparison with short-term exposures to microgravity within a range of minutes or seconds (see above). Similar observations were also reported by others. After 4 days in space, *Arabidopsis* plants exhibited only 27 transcripts which were at least twofold altered in their expression level (Correll et al. 2013). This might indicate that plants respond immediately to a microgravity environment but then adapt to the new situation on the longer run. Zhang and coworkers could identify only 45 proteins changed in expression after 14 days in space (same mission Zhang et al. 2015). Genes with prolonged changes in expression could, however, provide important information about the physiological needs after a few days in space. These include an upregulated group of genes which code for proteins that constitute the ribosomal complex within plastids. They are necessary for translation of mRNA. The upregulation of members of the mitochondrial electron transport chain, together with the upregulated expression of NADH dehydrogenase, could indicate an increased need for ATP. Interestingly, gene products involved in processes like the response to stress, protein degradation, or programmed cell death appeared not to be altered in expression. Together with the fact that there were no significant changes in gene expression detectable after 5 days of microgravity lets us suggest that at this stage the impact of a lack of gravitation on cell physiology was not too strong any more. The space environment per se, however, causes possibly an increased energy demand, as shown by the upregulation of respiratory components. This aspect should be taken into consideration when plants will be used to provide nutrients, oxygen, and energy on long-duration space missions.

*H*eat *s*hock *p*roteins (HSPs) dominate the group of transcripts which are reduced in amount (not shown). These proteins are involved in many forms of stress response. They enable the folding and membrane translocation of proteins and are thought to reconstitute the tertiary structure of proteins affected by stress events. They can increase the stress tolerance. A decreased expression should thus indicate a lower number of proteins affected in their structure (see also Salmi and Roux 2008; Manzano et al. 2012). There are, however, also reports on increased expression of HSPs (Paul et al. 2005, 2011, 2012; Zupanska et al. 2013; Hausmann et al. 2014).

A group of plant genes which are always affected by altered gravity are those involved in cell wall modification (Hoson et al. 2002, 2003; Wang et al. 2006; Correll et al. 2013; Nasir et al. 2014). This reflects the need for increased (hypergravity) or a reduced need for stability (microgravity). In our studies, expression of expansines (cell wall loosening activity) was increased (not shown). This might be the reason for the enhanced size of the microgravity-exposed cultures when compared to the 1 *g* controls.

2.3.5.9 Impact of the Non-microgravity-Related Spaceflight Conditions on Gene Expression

The availability of a 1 *g* reference centrifuge enabled us to screen for genes affected by non-microgravity-related spaceflight conditions, in that we compared expression data between µ*g* exposed and 1 *g* space with 1 *g* ground samples. This resulted in a considerable number of identical genes altered in mRNA abundance (573 genes; Fig. 2.26). We thus assume that this could be due to effects of spaceflight-related environmental conditions, including space radiation. Radiation measurements inside the capsule in a position close to our samples yielded a total dose of 5.9–8.1 mSV (millisieverts) and an equivalent dose of 370–510 µSV/d (data, Chinese authorities). This is considerably more compared to terrestrial conditions (3–6 µSV/d) and could be one of the reasons for the alterations in transcript levels, obviously not related to µ*g*. Zhang et al. also reported a greater effect on protein expression of non-µ*g* conditions (Zhang et al. 2015). Analysis showed that both experimental conditions (µ*g* and non-µ*g* spaceflight conditions) affected different biological processes. Overrepresented processes should not be regarded separately, as they are closely linked together within a plant cell. As mentioned already, the formation of *r*eactive *o*xygen *s*pecies (ROS) is one of the initial responses upon most kinds of stresses. The role of ROS in response to environmental changes can be deduced from alterations in gene products, involved in ROS production and turnover. Again we observed that many ROS-related genes are significantly regulated (Table 2.8). They comprise peroxidases, catalase, and a glutathione S-transferase. The increase in detoxification-related transcripts appears reasonable, as radiation in orbit consists of highly energetic (HZE) particles from interplanetary galactic sources or results from solar particle events, which could have an impact on cells (Horneck 1992, 1999; Baumstark-Khan et al. 2005; Kennedy 2014). It was shown that X-rays, γ-rays, protons, and heavy charged particles increased oxidative stress in different cell types, and countermeasures for space radiation effects are the use of antioxidants (Wan et al. 2003, 2005; Kennedy 2014). Similar responses are probable for plant cells. Therefore, the impact of long-term space radiation on the transcriptome of *Arabidopsis* should be investigated in ground-based studies in simulation test beds for the space environments (Horneck 1999).

In addition, a range of WRKY transcription factors and components of signaling chains (Ca^{2+}-dependent proteins, MAP kinases) were identified (Table 2.8). These responsive kinases are potentially also modulated by cytosolic fluctuations of H_2O_2 and can thus be part of signal transduction chains starting from hydrogen peroxide (for defense-related genes in tomato, see Orozco-Cárdenas et al. 2001). Thus, genes which are defense-, resistance-, and pathogen-related and which have been shown to respond to altered gravitation also respond to non-μ*g*-related spaceflight conditions.

2.3.6 *Protein Expression and Modulation*

As shown above, callus cell cultures of *Arabidopsis thaliana* respond to changes in gravitational field strengths by altered gene expression. Alterations in gene expression can result in altered protein patterns or in modifications of existing proteins. Protein expression is necessarily slower than changes in transcript levels and, with regard to the time axis, cannot be compared directly. Changes in protein modulation (phosphorylation, dephosphorylation), in contrast, can occur very rapidly. We thus tried to investigate both aspects.

Due to the methods available at the time, we started with two-dimensional SDS polyacrylamide gel electrophoresis (2-D SDS PAGE). In the respective studies, we investigated how different *g* conditions affect the proteome/phosphoproteome of A.t. cells (Barjaktarović et al. 2007, 2009a, b). For this purpose, callus cells were exposed to 8 *g* (centrifugation) and simulated microgravity (2-D clinorotation; 3-D random positioning) for up to 16 h. Extracts containing total soluble protein were subjected to 2-D SDS PAGE. Sypro Ruby®-stained gels were subjected to image analysis.

2.3.6.1 Hypergravity

After 2 h of hypergravity (8 *g*), 18 spots were upregulated, and 10 were downregulated. These spots/proteins were identified by correlating the ESI-MS/MS spectra with the NCBI nr-protein sequence database *Arabidopsis thaliana* as of August 06, 2007, using the MOWSE algorithm as implemented in the MS search engine MASCOT (Matrix Science Ltd. London, UK) (Cottrell and London 1999;

Barjaktarović et al. 2007). The identified proteins are involved in different stress responses/detoxification reactions (HSP91, methionine synthase, methionine sulfoxide reductase, aldehyde dehydrogenase, FQR1, aconitase, CR88), in protein folding (chaperones; HSP60, HSP91, calnexin, annexin, protein disulfide isomerase), regulation of gene expression (MBD10), cell division plate formation (CDC48), photomorphogenesis (AJH1, CR88), and metabolism [CPN60, alanine aminotransferase, lipoamide dehydrogenase, aconitase, aldolase, malate dehydrogenase (NADP)].

The amounts of the respective proteins fluctuated between 2 and 16 h (Barjaktarović et al. 2007) to a different extent.

2.3.6.2 2-D Clinorotation

To a considerable extent (20 out of 28 proteins), alterations in protein abundance after 2-D clinorotation are comparable to those under hypergravity (8 *g*/2-D; ratios after 2 h of treatment above 1.5 or below 0.5; Figs. 2.27a, b; Table 2.9). Differences are most obvious for HSP60 (assembly of multimeric proteins, activation of mitochondrial biogenesis) which is upregulated upon clinorotation while downregulated under centrifugation; the opposite was found for inorganic diphosphatase (energy supply for, e.g., vacuolar ATPases) and ATP-binding protein (Barjaktarović et al. 2007).

2.3.6.3 Random Positioning

RPM exposure caused considerably less changes in amounts of proteins. Only 8 out of the 28 proteins affected by hyper-*g* and 2-D were altered (Fig. 2.27c; Table 2.9). Generally, and with the exception of HSP91, responses were weaker in comparison with the other treatments. Some similarities in behavior existed for aminopeptidase, AJH1, MBD 10 (all 8 *g*, 2D), HSP60, alanine aminotransferase (both 2-D), HSP91, and protein disulfide isomerase (both 8 *g*). Only aldehyde dehydrogenase (detoxification; downregulated) and aconitase (tricarboxylic acid cycle; increase after initial downregulation) showed changes in abundance which were different from those of either 8 *g* or 2-D clinorotation.

Taken together the analyses show that the identified proteins were mainly stress-related and involved in detoxification of reactive oxygen species, signaling, and calcium binding. Surprisingly, centrifugation and clinorotation showed homologies which were not detected for random positioning. The protein data indicate that the response to simulation of weightlessness is different between clinorotation and random positioning.

Phosphoproteom

Two-dimensional SDS Polyacrylamide Gel Electrophoresis

Regulation/adaptation of metabolism to altered environmental conditions can rapidly be achieved by protein modulation. We thus extended these investigations to phosphorylated proteins (Barjaktarović et al. 2009a, b). Callus cell cultures of *A. thaliana* were exposed to hypergravity (8 *g*) and simulated weightlessness (random positioning) for up to 30 min, a sufficiently long exposure time which yielded reliable data. First changes, however, were visible as early as 10 min after start of treatment.

In comparison with 1 *g* controls, exposure to hypergravity resulted in 18 protein spots, and random positioning in 25, respectively, with increased/decreased signal intensity by at least twofold ($p < 0.05$). Only one spot (alanine aminotransferase) responded the same way under both treatments. After 30 min of RP, four spots appeared, which could not be detected in control samples. Out of the protein spots altered in phosphorylation, we were able to identify 24 from those responding to random positioning and 12 which responded to 8 *g*. These 12 proteins (8 *g*) were partly (5 out of 12) the same as those changed in expression after exposure to 2 h of hypergravity (Barjaktarović et al. 2007).

Figure 2.28 gives examples for 2-D PAGE separations of phosphorylated proteins after exposure to 8 *g* and RP and staining with Pro-Q Diamond (Barjaktarović et al. 2009b). According to function, the proteins altered in their degree of phosphorylation are again involved in ROS scavenging and detoxification (32%), primary metabolism (20.5%), general signaling (14.7%), protein translation and proteolysis (14.7%), and ion homeostasis (8.8%).

In addition to the data on gene expression, this gives another support for the notion that changes in gravitational fields induce the production of ROS. Our data further indicate that responses toward RP are more by posttranslational protein

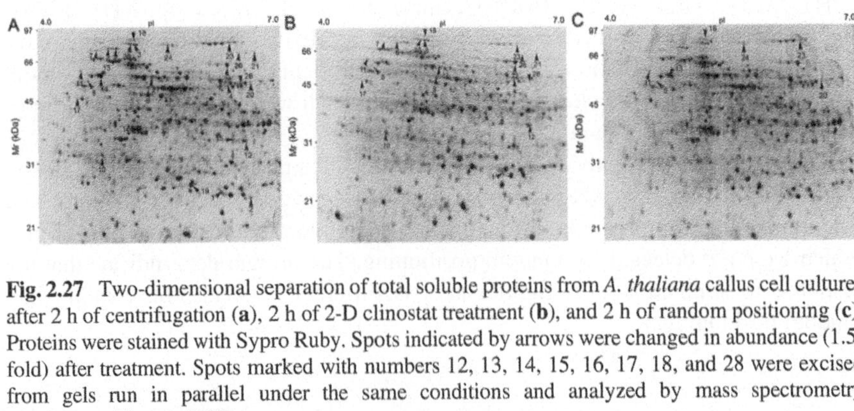

Fig. 2.27 Two-dimensional separation of total soluble proteins from *A. thaliana* callus cell cultures after 2 h of centrifugation (**a**), 2 h of 2-D clinostat treatment (**b**), and 2 h of random positioning (**c**). Proteins were stained with Sypro Ruby. Spots indicated by arrows were changed in abundance (1.5-fold) after treatment. Spots marked with numbers 12, 13, 14, 15, 16, 17, 18, and 28 were excised from gels run in parallel under the same conditions and analyzed by mass spectrometry (Barjaktarović et al. 2007)

Table 2.9 Functionally annotated proteins with significant changes in abundance after 2 h of exposure to 8 *g*, 2-D clinostatting and random positioning

Spot no	*Arabidopsis* protein name	Biological process	8 *g*/ 1 *g*	2-D/ 1 *g*	3-D/ 1 *g*
1	FQR1	Electron transport, auxin stimulus	1.5	2.1	1.0
2	Protein-methionine-S-oxide reductase	Protein binding, protein metabolic process	1.6	1.0	1.0
3	CDC48	Phosphorylation	2.7	1.5	1.0
4	ALDH2B7; aldehyde dehydrogenase	Metabolic process	2.8	2.0	1.0
5	ATP binding	Protein folding	2.1	1.0	1.0
6	Unknown protein	Unknown	1.5	1.5	1.0
7	CR88	De-etiolation	3.3	2.6	1.0
8	AJH1	Response to auxin, photomorphogenesis, protein deneddylation	2.6	1.5	1.6
9	ALAAT1 (alanine aminotransferase)	L-alanine catabolic process, response to hypoxia	1.5	1.5	1.4
10	Inorganic pyrophosphatase	Metabolic process	1.5	0.6	1.0
11	MBD10; DNA binding	DNA binding	1.5	1.5	1.5
12	Malate dehydrogenase	Tricarboxylic acid cycle, malate metabolism	1.8	1.5	1.4
13	ATPDIL1-2; protein disulfide isomerase	Protein disulfide isomerase, electron transporter	1.9	1.4	1.6
14	CPN60A	Protein binding, ATP binding	1.8	1.3	1.0
15	Fructose bisphosphate aldolase	Pentose phosphate shunt	2.1	2.9	1.0
16	ANNAT1 (annexin Arabidopsis1)	Response to oxidative stress	2.1	3.2	1.5
17	Calcium ion binding (calnexin)		2.0	1.5	1.0
18	HSP 91	Protein folding, response to heat	2.0	2.0	2.9
19	Glutathione transferase	Toxin catabolism, response to oxidative stress	0.3	1.0	1.0
20	Lipoamide dehydrogenase	Dihydrolipoyl dehydrogenase, oxidoreductase, FAD binding	0.6	1.0	0.5
21	Methionine synthase	Methionine biosynthesis	0.3	0.3	1.0
22	Stress inducible protein	Stress response	0.3	0.5	1.0
23	Aconitase	Response to oxidative stress	0.5	0.7	0.5
24	Putative aminopeptidase	Proteolysis	0.5	0.5	0.6
25	ATP binding/hydrolase	Unknown	0.4	0.4	1.0
26	Methionine synthase	Methionine biosynthesis	0.6	0.6	1.0
27	HSP 60	Response to heat, protein assembly	0.4	2.0	1.4
28	ATP synthase subunit alpha. mitochondrial	ATP synthesis-coupled transport	0.3	0.4	1.0

8 *g*/control, 2-D/control, 3-D/control: ratio of amount of identified protein of treated sample and stationary control (Barjaktarović et al. 2007)

modulation (most changes in the degree of phosphorylation occur under RP treatment) than by protein expression (hypergravity).

2.3.6.4 Protein Labeling

SILAC, following 2-D gel electrophoresis, was a large step forward in protein analysis. SILAC (stable isotope labeling by amino acids in cell culture) is now a widespread method for metabolic labeling of cells and tissues in quantitative proteomics.

In a typical SILAC experiment (Ong et al. 2002; Ong and Mann 2006), cells are grown in a defined medium complemented with certain essential amino acids, usually arginine and lysine, containing naturally occurring atoms (the "light" culture) or a specific number of their stable isotope counterparts (the "heavy" culture). After full incorporation of the labeled amino acid, the light and heavy cultures are subjected to different perturbations, harvested, and combined. Peptides resulting from enzymatic digest of the combined cell lysate are then detected by mass spectrometry (MS) in form of ion pairs (doublets), and their ratios reflect the relative changes in protein abundance between the light and the heavy culture (Schütz et al. 2011). For SILAC labeling, Lys-4 (4.4.5.5.-$^{2}H_4$) or Lys-8 ($^{13}C_6{}^{15}N_2$) were added daily for 12 days before an experiment.

In addition we carried out ^{15}N labeling with ammonium nitrate. For this purpose, a "heavy" culture medium was prepared without amino acids and in which potassium nitrate and ammonium nitrate were replaced by half of the standard amount of ^{15}N containing potassium nitrate and ammonium nitrate (Euriso-top, Saint-Aubin Cedex, France).

Results from ^{15}N labeled A.t. callus cultures, harvested during parabolic flights after the microgravity phase, are given for proteins altered in the degree of phosphorylation in relation to their function in Tables 2.10 and 2.11 (Karl 2015; Karl, Hausmann, Schulze, Macek, Hampp; unpublished results).

With regard to metabolism, the ^{15}N data reveals alterations for many proteins, with a focus on primary metabolism, and here especially carbohydrate metabolism (Table 2.10). In this context the double-function protein, 6-phosphofructo-2-kinase/fructose-2,6-bisphosphatase is of interest because it constitutes an important regulator of gluconeogenesis/glycolysis (compare changes in concentration of its metabolic regulator, $F_{26}BP$ (Fig. 2.7). Together with other enzymes of these pathways, and the citric acid cycle, such as aldolase, G6PDH, pyruvate kinase, PEP carboxylase, NADP malic enzyme, and succinate dehydrogenase, regulation of glycolysis and amino acid synthesis appears to be important under transition to microgravity.

In Table 2.11, proteins (27) are compiled which are stress-related. Again, some of them are involved in the detoxification of reactive oxygen species (glutathione S-transferase; glutathione peroxidase).

In addition, a considerable amount of proteins was identified (not shown) which regulate the state of phosphorylation of other proteins (67). These constitute mainly

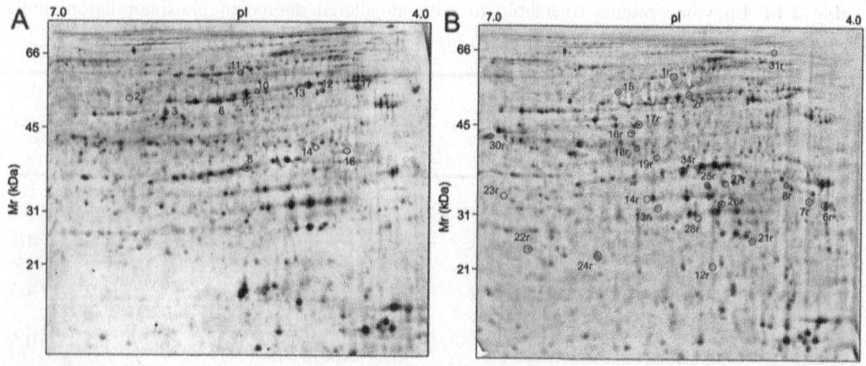

Fig. 2.28 Two-dimensional separation of phosphoproteins from *A. thaliana* callus cell cultures. The gels were stained with Pro-Q Diamond. Indicated spots are those with increased/decreased signal intensity (PD Quest) by at least twofold ($p < 0.05$) after 30 min of 8 *g* (**a**) and 30 min of random positioning treatment (**b**). The ranges of pI and molecular masses are indicated (Barjaktarović et al. 2009b)

protein kinases and phosphatases. Some of them are involved in Ca^{2+}- or ROS-dependent pathways (calmodulin, CDPKs).

Regulators of transcription were also subject to control by altered phosphorylation (27 proteins). Some of these proteins are Ca^{2+}-dependent.

Other groups of modified proteins are involved in RNA binding, processing, and translation (81), transport (60; clathrin interactors, clathrin assembly proteins, intracellular protein, or membrane transport), protein turnover (22; protein degradation via the ubiquitin pathway and the proteasome), and cytoskeleton-related proteins (motor proteins, linkers, or structural components of the cytoskeleton and cytoskeleton-associated proteins).

For most of the proteins which change their degree of phosphorylation, it is not yet known, how this affects their activity or their interaction with other proteins. Most of the changes detected show an increase in the amount of the phosphorylated protein in relation to the 1 *g* control. For proteins such as aldehyde dehydrogenase, CPN60B (CPN60), CPN60A (HSP60), or alanine aminotransferase (30 min, 8 *g*), this should mean increased activity, because these proteins appeared also to be increased in amount after 2 h of exposure to hypergravity (Barjaktarović et al. 2007). But there are also quite a few with a decrease in phosphorylation.

2.3.6.5 Evaluation of Simulation of Microgravity

Interestingly, the responses discussed so far are primarily related to centrifugation and 2-D clinorotation, while samples exposed to random positioning show weaker effects. The unexpected similarities between hypergravity and clinorotation could be due to either residual gravity or permanent stimulation due to rotation. According to calculations, our cell cultures experienced only 0.0016 *g* at the highest (60 rpm;

Table 2.10 Enzyme proteins (metabolism) with an altered degree of phosphorylation under microgravity (^{15}N labeled)

Protein	Recommended Name	Modified Sequence	Exp. 1 Ratio µg/1g	Exp. 2 Ratio µg/1g
Q8W108	1,2-dihydroxy-3-keto-5-methylthiopentene dioxygenase 3	_VNEGGVIDAS(ph)A_	2.44	2.45
P54887	Delta1-pyrroline-5-carboxylate synthase 1	_YRQLVNS(ph)SFADLQK_	4.83	4.18
P54887	Delta1-pyrroline-5-carboxylate synthase 1	_QLVNSS(ph)FADLQK_	6.23	4.52
Q42521	Glutamate decarboxylase 1	_SESNS(ph)DNLM(ox)VTVK_	0.51	0.58
P49572	Indole-3-glycerol phosphate synthase, chloroplastic	_ESLAVSSS(ph)SVEDK_	0.90	0.99
Q9MB58	6-phosphofructo-2-kinase/fructose-2,6-bisphosphatase	_SLS(ph)ASSFLIDTK_	3.10	1.97
C0Z2J1	Adenine nucleotide alpha hydrolases-like protein	_SGDS(ph)GGENLMDSDK_	4.83	1.26
P31166	Adenine phosphoribosyltransferase 1, chloroplastic	_(ac)AT(ph)EDVQDPR_	8.51	1.91
P06525	Alcohol dehydrogenase class-P	_IIGVDFNS(ph)K_	2.22	1.26
P06525	Alcohol dehydrogenase class-P	_(ac)ST(ph)TGQIIR_	0.07	0.09
Q9SJQ9	Fructose-bisphosphate aldolase	_LGEGAAES(ph)LHVK_	0.35	0.52
Q9LF98	Fructose-bisphosphate aldolase	_LGDGAAES(ph)LHVK_	1.11	2.36
Q9FJI5	Glucose-6-phosphate 1-dehydrogenase	_(ac)GS(ph)GQWHVEK_	0.20	0.47
Q9LYG3	NADP-dependent malic enzyme 2	_(ac)GS(ph)TPTDLPGEDVADNR_	0.13	0.66
Q84VW9	Phosphoenolpyruvate carboxylase 3	_S(ph)AQELVK_	9.69	2.10
Q9FLH8	Probable fructokinase-7	_(ac)GEDAIS(ph)GNLK_	0.33	0.25
Q9FFP6	Pyruvate kinase	_(ac)S(ph)NIDIEGILK_	4.66	1.72
O82663	Succinate dehydrogenase [ubiquinone] flavoprotein subunit 1, mitochondrial	_NSNGS(ph)LPTSTIR_	2.92	1.09
O04905	Uridine monophosphate/cytidine monophosphate kinase 3	_(ac)GS(ph)VDAANGSGK_	1.13	1.44

(continued)

Table 2.10 (continued)

Q9SW48	Probable alkaline/neutral invertase B	_SLTELTGS(ph)PQLR_	2.11	1.08
Q9SN95	UDP-glucuronic acid decarboxylase 5	_QT(ph)SPKPPPSPSPLR_	2.08	4.87
Q9LD43	Acetyl-coenzyme A carboxylase carboxyl transferase subunit alpha, chloroplastic	_ELAAEES(ph)DGSVK_	2.95	5.03
B9DGD6	Acetyl-coenzyme A synthetase, chloroplastic/glyoxysomal	_HVES(ph)MSQLPSGAGK_	4.69	3.36
B9DGD6	Acetyl-coenzyme A synthetase, chloroplastic/glyoxysomal	_HVES(ph)M(ox)SQLPSGAGK_	5.00	2.20
Q8L5Y9	Pantothenate kinase 2	_S(ph)GSRPQLDLSK_	0.70	0.48
Q9SF47	Phosphatidate phosphatase PAH1	_FYDFQDDPPS(ph)PTSEYGSAR_	0.75	0.57
Q8L9X1	Phospholipase/carboxylesterase family protein	_(ac)S(ph)ISGAAVGS(ph)GR_	0.44	0.17
O80452	AMP deaminase	_TGS(ph)FVRPIS(ph)PK_	1.68	3.84
Q9M389	Embryonic abundant protein-like	_FESDQVLS(ph)NK_	2.26	1.17
Q9SQR4	NAD(P)-binding Rossmann-fold superfamily protein	_(ac)S(ph)THSSISQPPLPLAGR_	7.46	7.46
P11035	Nitrate reductase [NADH] 2	_SVS(ph)TPFMNTTAK_	2.57	1.50
P32962	Nitrilase 2	_(ac)STS(ph)ENTPFNGVASSTIVR_	0.61	0.56
Q9SRQ6	Non-specific phospholipase C3	_(ac)VEETS(ph)SGGGSSASPIK_	1.54	2.87
Q84WW3	Phosphoglycerate mutase-like protein	_QGS(ph)GIIGTFGQSEELR_	2.27	2.59
Q0WW53	Phosphoglycerate mutase-like protein	_VAS(ph)VLSQHRK_	5.40	6.88
Q9FVW3	Putative methylesterase 14, chloroplastic	_TLS(ph)DPFS(ph)NGK_	2.84	1.97
Q9FVW3	Putative methylesterase 14, chloroplastic	_TLS(ph)DPFSNGK_	1.02	6.36
Q9ZVC3	S-adenosyl-L-methionine-dependent methyltransferase-like protein	_VDNDESTES(ph)QSFPVR_	1.77	2.65
Q9LFF9	Soluble inorganic pyrophosphatase 4	_HVS(ph)LPTLNER_	4.48	1.73

Karl 2015; Karl, Hausmann, Schulze, Macek, Hampp unpublished

4 mm radius), which is below the gravity-response threshold (e.g., 0.16 *g* for *Euglena* cells (Häder and Hemmersbach 1997) and 0.05 *g* for the displacement of statoliths in *Chara* rhizoids (Limbach et al. 2005). Thus, it is at least similarly likely that permanent stimulation is the cause for similarities in the stress response. In contrast, the RPM-specific responses (aldehyde dehydrogenase (detoxification), downregulated, and aconitase (tricarboxylic acid cycle) increase after initial downregulation) could be interpreted as a decrease in ROS stress in comparison with 1 *g* samples [there is obviously always a basic ROS load also in unstressed cells

Table 2.11 Enzyme proteins (stress response and stress-related reactions) with an altered degree of phosphorylation under microgravity (^{15}N labeled)

Protein	Recommended Name	Modified Sequence	Exp. 1 Ratio µg/1g	Exp. 2 Ratio µg/1g
Q9LHL7	Aha1 domain-containing protein	_S(ph)GAAPVVSAAAVESK_	0.21	0.72
Q945Q1	Cysteine proteinase inhibitor 1	_(ac)ADQQAGT(ph)IVGGVR_	1.82	1.61
Q9LTM3	Cytochrome P450 71B20	_S(ph)PVDLSK_	8.05	8.25
O48832	Dehydrin ERD7	_EISHDS(ph)S(ph)DEEDGDGGK_	0.13	0.34
P42761	Dehydrin ERD10	_TQIS(ph)EPES(ph)FVAK_	0.04	0.76
P42760	Dehydrin ERD10	_VATEESS(ph)APEIK_	0.60	0.61
P42759	Dehydrin ERD10	_VATEES(ph)S(ph)APEIK_	0.06	0.15
P42763	Dehydrin ERD14	_VHIS(ph)EPEPEVK_	0.44	0.56
P46422	Glutathione S-transferase F2	_YLFSAESVS(ph)QR_	3.71	3.65
P42760	Glutathione S-transferase F6	_GNNLLST(ph)GK_	0.08	0.80
Q9SRY5	Glutathione S-transferase F7	_GNQLVSLGS(ph)K_	7.04	11.83
Q8H121	Glutathione S-transferase family protein	_SAVDETSDS(ph)GAFQR_	1.11	1.05
Q9XIE3	17.6 kDa class I heat shock protein 1	_(ac)S(ph)LIPSFFGNNRR_	2.22	3.97
Q8GUM2	Heat shock 70 kDa protein 9, mitochondrial	_SSGGLS(ph)DDEINR_	5.80	1.42
P55737	Heat shock protein 81-2	_EIS(ph)DDEEEEEK_	0.30	0.88
P27323	Heat shock protein 90-1	_EIS(ph)DDEDEDEPK_	9.40	6.95
P27323	Heat shock protein 90-1	_EIS(ph)DDEDEDEPKK_	5.89	8.01
F4JHJ0	Hsp20-like chaperone	_APAAEETTS(ph)VKEDK_	0.51	0.34
Q5XEP2	Hsp70-Hsp90 organizing protein 2	_ANRGDLT(ph)PEELK_	1.10	1.04
Q5XEP2	Hsp70-Hsp90 organizing protein 2	_NPS(ph)NLNLYLQDQR_	6.66	8.59
O04922	Probable glutathione peroxidase 2	_VLQRYS(ph)PR_	10.62	16.92
Q9FPF0A	Protein DJ-1 homolog	_S(ph)LENM(ox)VK_	4.98	12.01
F4I893	Protein ILITYHIA	_ALLEGGS(ph)DDEGASTEAQGR_	6.30	3.84
Q9FIJ0	Respiratory burst oxidase homolog protein D	_ILSQM(ox)LS(ph)QK_	1.89	12.48
P18612	Stress-induced protein KIN1	_S(ph)VSDAAAGGVNFVK_	0.87	0.88
P31169	Stress-induced protein KIN2	_DAAAAAGAS(ph)AQQAGK_	0.42	0.75
Q9M0B2	Tetratricopeptide repeat domain-containing protein	_QVDGDDS(ph)DGFETASER_	2.61	2.40

Karl 2015; Karl, Hausmann, Schulze, Macek, Hampp unpublished which could regulate the redox state (Halliwell 2006)]. Probably, under RP, changes in position/orientation are so fast that the cells "forget" about the gravity vector.

As far as comparable data exists, RPM treatments show similarities to real microgravity as experienced in sounding rocket experiments (own data) or space shuttle experiments (Marco et al. 2006; Barjaktarović et al. 2007). It is thus tempting to conclude that earth gravitation could impose some basic stress response which is relieved by reducing the gravitational field.

2.3.6.6 How We Think that a Gravity-Related Line of Plant Responses Works: Secondary Messengers, Genes, and Proteins

ROS Scavenging and Detoxifying Enzymes

Extracellular signals such as resulting from biotic or abiotic stresses are recognized by membrane-located receptors or sensors and are translated into the intracellular formation of signal mediators. Reactive oxygen species form an important group of such intermediates. Biotic as well as abiotic stresses induce a reduction of molecular oxygen by membrane-bound NADPH oxidase, giving rise to superoxide radicals. These are to some degree detoxified by discrimination into oxygen and hydrogen peroxide (superoxide dismutase). The latter product is also toxic but serves as an important signal for gene expression, leading either to programmed cell death or to the production of radical scavenging proteins. Balancing of production and detoxification of ROS is thus important for maintaining cellular functions. For scavenging, plants possess a large number of antioxidant enzymes (Mittler et al. 2004). Upon exposure to environmental stimuli and oxidative stress, transcripts of enzymes involved in ROS scavenging are induced. However, for the purpose of signaling, where small fluctuations of ROS are enough to activate transduction pathways, ROS production and scavenging enzymes are likely to be regulated by posttranslational modifications, especially in the seconds range.

We found the following ones to be altered in the degree of phosphorylation within 30 min of exposure: thioredoxin reductase, glutathione peroxidase, ascorbate peroxidase, ATP sulfurylase, alcohol dehydrogenase, P23 co-chaperone, and HSP83 (all random positioning), protein disulfide isomerase/aldehyde dehydrogenase, GrpE-like protein, and the chaperones CNP60 A and B (all 8 *g*).

Interestingly, some of these proteins increased in total amount starting from 2 h of exposure to hypergravity (Barjaktarović et al. 2007). Thus, modulation is a first and fast response, followed by an increase in amount of these ROS-related stress proteins.

Signal Transduction

Five proteins which are well known to function in signal transduction pathways were identified. 14-3-3-like protein, the Rab GTPase homolog A4a, and myo-inositol monophosphatase were differentially phosphorylated after 30 min of exposure to random positioning. The Ran-binding protein 1a and the putative WD-40 repeat protein responded to 30 min of 8 *g*.

14-3-3-like proteins play important roles in many pathways that are regulated by phosphorylation. We do, however, not know what the target proteins of this 14-3-3 protein are and how many of the respective genes are expressed in cell cultures. Thus, it is not yet possible to evaluate the impact of this phosphorylation on signaling.

Inositol 1,4,5-triphosphate also plays a critical role in gravity signaling (Perera et al. 2006). *Myo*-inositol monophosphatase catalyzes the dephosphorylation of *myo*-inositol to form inositol, which is a precursor of two important phosphoinositol (PI) signaling messengers, diacylglycerol, and myo-inositol-1,4,5-triphosphate (IP3). Diacylglycerol activates protein kinase C which in turn regulates many proteins by phosphorylation. In addition, IP3 binds to Ca^{2+} channels to release calcium from intracellular stores.

Rab proteins are GTP-binding proteins belonging to the Ras superfamily and are involved in membrane trafficking pathways. The gravitational signal should thus also interact with transport processes.

WD40 repeats (beta-transducin repeats) are proteins with conserved domains of 40 amino acid motifs, terminating with Trp-Asp (W-D) and Gly-His dipeptides with 4 to 16 repeating units. A common function of WD40 repeats is the facilitation of protein-protein interactions, thereby forming regulatory complexes with, e.g., transcription factors (Ramsay and Glover 2005). We could show that a gravitation-related expression of transcription factors can occur within minutes (Martzivanou et al. 2006; Babbick et al. 2007).

Metabolism

It is known for long that the alteration of gravitational forces results in metabolic responses in plant cells (see above and Obenland and Brown 1994; Vasilenko and Popova 1996; Hampp et al. 1992, 1997; Wang et al. 2006). These include alterations in the energy and redox state of the cells, as well as in carbohydrate metabolism. Therefore, it is not surprising that many of the enzymes which are differentially stained with Pro-Q Diamond after 30 min of exposure to *random positioning* are involved in primary metabolism (glycolysis: fructose-1,6-bisphosphate aldolase, triose phosphate isomerase, phosphoglycerate mutase). Additionally, 6-phosphogluconolactonase and alanine aminotransferase (alanine catabolism) were found differentially phosphorylated. Fructose-1,6-bisphosphate aldolase and alanine aminotransferase were dephosphorylated compared to the stationary control, while triose phosphate isomerase, 6-phosphogluconolactonase, and phosphoglycerate mutase were phosphorylated. The increased ^{15}N-labeling of 6-phosphofructo-2-kinase/fructose-2,6-bisphosphatase under microgravity (parabolic flights) is in this context especially of interest, because it constitutes an important regulator of gluconeogenesis/glycolysis.

6-Phosphogluconolactonase is the second enzyme of oxidative pentose phosphate pathway (OPPP). The OPPP provides important intermediates for nucleotide biosynthesis and NADPH for enzymes regulating the redox state of the cell (glutathione/glutaredoxine and thioredoxin systems) (Grant 2008).

The responses to hypergravity were different. First, we found less glycolytic enzyme proteins altered in phosphorylation. Second, the amount of phosphorylated alanine aminotransferase was decreased as shown for RP, while phosphoglycerate mutase, in contrast to RP, was also less phosphorylated. Third, enolase showed decreased phosphorylation. Enolase is located at the converging point between glycolysis and gluconeogenesis and catalyzes the conversion of 2-phosphoglycerate to phospho*enol*pyruvate. It is activated by dephosphorylation (Lal et al. 1998; Forsthoefel et al. 1995), which infers an upregulation of this step in glycolysis. In our previous studies, we also found evidence for increased rates of glycolysis and anaplerosis under altered *g* conditions (Hampp et al. 2001; Martzivanou and Hampp 2003).

Protein Translation and Proteolysis

We have identified two ribosomal proteins, 60S acidic ribosomal protein P0-B (random positioning) and 60S acidic ribosomal protein P0-C (8 *g*) with increased phosphorylation after treatment. Ribosomal P-proteins are acidic proteins that interact with tRNA, mRNA, and elongation factors during translation. Phosphorylation of these proteins leads to increased (selective) translation activity (Bailey-Serres et al. 1997; Aguilar et al. 1998). This is in accordance with increased gene expression which can start within minutes after altering the gravitational field (e.g., transcription factors, kinases, etc.; Moseyko et al. 2002; Kimbrough et al. 2004; Martzivanou et al. 2006) and with increased amounts of ribosomal proteins upon exposure to clinorotation (Wang et al. 2006). Additionally, two subunits of proteasome 20S (PBD1 and PAF1) exhibited increased phosphorylation in response to random positioning and a metalloendopeptidase to 8 *g*. Proteasomes are important for protein turnover, and their phosphorylation regulates activity and structure (Umeda et al. 1997; Bose et al. 2004).

Cell Wall Biosynthesis

Reversibly glycosylated polypeptides (RGPs) are highly conserved plant proteins, probably involved in plant polysaccharide synthesis (Delgado et al. 1998). Self-glycosylation of RGPs increases their ability to become a part of protein complexes (De Pino et al. 2007). On the other hand, phosphorylation reduces self-glycosylation by decreasing their affinity to sugars. RGP1 responded to random positioning by an increase in the phosphorylation status. RGPs are also differentially expressed after long-term exposure to hypergravity (own data, not shown) or in response to clinorotation (Wang et al. 2006). There is also a lot of other data, showing that the expression of cell wall-related proteins is well affected by altered gravitational fields (Martzivanou and Hampp 2003; Hoson and Soga 2003; Hoson et al. 2003; Yoshioka et al. 2003; Soga 2013).

Ion homeostasis

Protons are implicated in plant signal transduction as second messengers (Roos et al. 1998). One of the early responses to gravistimulation is alkalinization of the cytosol (Fasano et al. 2001; Johannes et al. 2001). pH oscillations could be a result of the activation of plasma membrane or vacuolar H^+-ATPases (Perrin et al. 2005). Vacuolar (V) ATPase appears as a phosphorylated protein upon exposure to random positioning. The spot was not detectable in control samples. In barley, subunits A and B of V-ATPase are targets of 14-3-3-proteins (see above). 14-3-3 binding leads to the activation of the V-ATPase (Klychnikov et al. 2006). Thus V-ATPase could play a role in proton homeostasis and early signaling as a response to changes in the gravitational force.

Band 7 protein, which probably also functions in ion homeostasis, responded after both treatments (RP increased phosphorylation, 8 *g* decreased phosphorylation; distinct spots). In a study by Davletova et al. (Davletova et al. 2005), the transcript level of band 7 protein was upregulated upon oxidative and light stress. In our proteome study, the expression of this protein was also increased upon long-term exposure to 8 *g* (data not shown).

2.4 Conclusions

We present an analysis of second messengers in combination with downstream responses at the level of gene and protein expression and protein modulation. The data show that in a sequence of changing gravitational field strengths, the secondary messengers Ca^{2+} and hydrogen peroxide respond to a lack of gravity by a rapid increase. The rise of hydrogen peroxide production is most probably triggered by Ca^{2+}. Resulting changes in protein modulation indicate the initiation of signaling chains, as well as fast adaptations in primary metabolism. Concomitant changes in transcripts are connected with Ca^{2+}-related steps and fine tuning of levels of reactive oxygen species (Fig. 2.29).

Trials to compare molecular responses under simulated microgravity with true microgravity are not consistent.

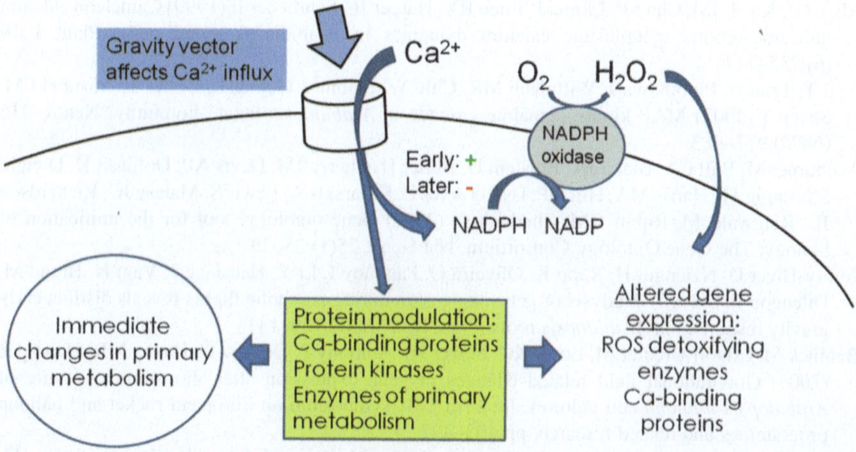

Fig. 2.29 Schematic presentation of assumed early events upon the onset of microgravity. Ca^{2+} and H_2O_2 rise in parallel. The link between both second messengers could be the activation by Ca^{2+} of a plasma membrane-located NADPH oxidase. This enzyme could be inhibited by a further rise in Ca^{2+}. Both Ca^{2+} and H_2O_2 interfere with gene expression and protein modulation. This can result in a fast response of metabolic pathways possessing key enzymes that are regulated by phosphorylation but also in gene expression through the modification of transcription factors (Hausmann et al. 2014)

- RPM-induced changes in gene expression resembled those of magnetic levitation.
- Changes in gene expression due to two-dimensional clinorotation were similar to those caused by hypergravity.
- Changes in protein modulation caused by hypergravity showed homologies with two-dimensional clinorotation.
- Such homologies were not detected for random positioning.

Thus, responses to 2-D clinorotation and hypergravity are closer to each other than to clinorotation and random positioning. The simulation of true microgravity is probably not completely possible. However, it can be concluded from our data that RPM and magnetic levitation are more suitable to simulate microgravity than 2-D clinorotation.

The return to the nearly-1 *g*-behavior after extended periods of exposure (e.g., 5 days in space) lets us assume that plant cells are highly adaptive to changing environmental conditions.

References

Aguilar R, Montoya L, de Jiménez ES (1998) Synthesis and phosphorylation of maize acidic ribosomal proteins implications in translational regulation. Plant Physiol 116(1):379–385

Allen GJ, Kwak JM, Chu SP, Llopis J, Tsien RY, Harper JF, Schroeder JI (1999) Cameleon calcium indicator reports cytoplasmic calcium dynamics in Arabidopsis guard cells. Plant J 19 (6):735–747

Asai T, Tena G, Plotnikova J, Willmann MR, Chiu WL, Gomez-Gomez L, Boller T, Ausubel FM, Sheen J (2002) MAP kinase signaling cascade in *Arabidopsis* innate immunity. Nature 415 (6875):977–983

Ashburner M, Ball CA, Blake JA, Botstein D, Butler H, Cherry JM, Davis AP, Dolinski K, Dwight SS, Eppig JT, Harris MA, Hill DP, Issel-Tarver L, Kasarskis A, Lewis S, Matese JC, Richardson JE, Ringwald M, Rubin GM, Sherlock G (2000) Gene ontology: tool for the unification of biology. The Gene Ontology Consortium. Nat Genet 25(1):25–29

Aubry-Hivet D, Nziengui H, Rapp K, Oliveira O, Paponov I, Li Y, Hauslage J, Vagt N, Braun M, Ditengou FA (2014) Analysis of gene expression during parabolic flights reveals distinct early gravity responses in *Arabidopsis* roots. Plant Biol 16(s1):129–141

Babbick M, Cogoli-Greuter M, Lowe KC, Power JB, Anthony P, Dijkstra C, Davey MR, Hampp R (2005) Gravitational field related changes in gene expression after short-term exposure of *Arabidopsis thaliana* cell cultures. In: 17th ESA symposium on European rocket and balloon programmes and related research, pp 493–497

Babbick M, Dijkstra C, Larkin O, Anthony P, Davey M, Power J, Lowe K, Cogoli-Greuter M, Hampp R (2007) Expression of transcription factors after short-term exposure of *Arabidopsis thaliana* cell cultures to hypergravity and simulated microgravity (2-D/3-D clinorotation, magnetic levitation). Adv Space Res 39(7):1182–1189

Bailey-Serres J, Vangala S, Szick K, Lee C-HK (1997) Acidic phosphoprotein complex of the 60S ribosomal subunit of maize seedling roots (components and changes in response to flooding). Plant Physiol 114(4):1293–1305

Baluska F, Hasenstein KH (1997) Root cytoskeleton: its role in perception of and response to gravity. Planta 203(Suppl):S69–S78

Barjaktarović Ž, Nordheim A, Lamkemeyer T, Fladerer C, Madlung J, Hampp R (2007) Time-course of changes in amounts of specific proteins upon exposure to hyper-g, 2-D clinorotation, and 3-D random positioning of *Arabidopsis* cell cultures. J Exp Bot 58(15–16):4357–4363

Barjaktarović Ž, Babbick M, Nordheim A, Lamkemeyer T, Magel E, Hampp R (2009a) Alterations in protein expression of *Arabidopsis thaliana* cell cultures during hyper-and simulated micro-gravity. Microgravity Sci Technol 21(1–2):191–196

Barjaktarović Ž, Schütz W, Madlung J, Fladerer C, Nordheim A, Hampp R (2009b) Changes in the effective gravitational field strength affect the state of phosphorylation of stress-related proteins in callus cultures of *Arabidopsis thaliana*. J Exp Bot 60(3):779–789

Batistic O, Kudla J (2012) Analysis of calcium signaling pathways in plants. Biochim Biophys Acta 1820:1283–1293

Baumstark-Khan C, Hellweg CE, Arenz A, Meier MM (2005) Cellular monitoring of the nuclear factor κB pathway for assessment of space environmental radiation. Radiat Res 164(4):527–530

Behera S, Krebs M, Loro G, Schumacher K, Costa A, Kudla J (2013) Ca^{2+} imaging in plants using genetically encoded yellow cameleon Ca^{2+} indicators. Cold Spring Harb Protoc 2013 (8):700–703

Belousov VV, Fradkov AF, Lukyanov KA, Staroverov DB, Shakhbazov KS, Terskikh AV, Lukyanov S (2006) Genetically encoded fluorescent indicator for intracellular hydrogen peroxide. Nat Meth 3(4):281–286

Bose S, Stratford FL, Broadfoot KI, Mason GG, Rivett AJ (2004) Phosphorylation of 20S proteasome alpha subunit C8 (alpha7) stabilizes the 26S proteasome and plays a role in the regulation of proteasome complexes by gamma-interferon. Biochem J 378(Pt 1):177–184

Braun M, Limbach C (2006) Rhizoids and protonemata of characean algae: model cells for research on polarized growth and plant gravity sensing. Protoplasma 229(2–4):133–142

Catherall A, Lopez-Alcaraz P, Benedict K, King P, Eaves L (2005) Cryogenically enhanced magneto-Archimedes levitation. New J Phys 7(1):118

Chen R, Guan C, Boonsirichai K, Masson PH (2002) Complex physiological and molecular processes underlying root gravitropism. Plant Mol Biol 49(3–4):305–317

Correll MJ, Pyle TP, Millar KD, Sun Y, Yao J, Edelmann RE, Kiss JZ (2013) Transcriptome analyses of *Arabidopsis thaliana* seedlings grown in space: implications for gravity-responsive genes. Planta 238(3):519–533

Costa A, Drago I, Behera S, Zottini M, Pizzo P, Schroeder JI, Pozzan T, Lo Schiavo F (2010) H_2O_2 in plant peroxisomes: an in vivo analysis uncovers a Ca(2+)-dependent scavenging system. Plant J 62(5):760–772

Cottrell JS, London U (1999) Probability-based protein identification by searching sequence databases using mass spectrometry data. Electrophoresis 20(18):3551–3567

Davletova S, Schlauch K, Coutu J, Mittler R (2005) The zinc-finger protein Zat12 plays a central role in reactive oxygen and abiotic stress signaling in Arabidopsis. Plant Physiol 139 (2):847–856

De Pino V, Borán M, Norambuena L, González M, Reyes F, Orellana A, Moreno S (2007) Complex formation regulates the glycosylation of the reversibly glycosylated polypeptide. Planta 226 (2):335–345

Delgado IJ, Wang Z, de Rocher A, Keegstra K, Raikhel NV (1998) Cloning and characterization of AtRGP1 a reversibly autoglycosylated Arabidopsis protein implicated in cell wall biosynthesis. Plant Physiol 116(4):1339–1350

Dodd AN, Kudla J, Sanders D (2010) The language of calcium signaling. Annu Rev Plant Biol 61:593–620

Eulgem T, Rushton PJ, Robatzek S, Somssich IE (2000) The WRKY superfamily of plant transcription factors. Trends Plant Sci 5(5):199–206

Fasano JM, Swanson SJ, Blancaflor EB, Dowd PE, Kao T-h, Gilroy S (2001) Changes in root cap pH are required for the gravity response of the Arabidopsis root. Plant Cell 13(4):907–921

Fasano JM, Massa GD, Gilroy S (2002) Ionic signaling in plant responses to gravity and touch. J Plant Growth Regul 21(2):71–88

Feige B, Gimmler H, Jeschke WD, Simonis W (1969) Eine Methode zur dünnschicht-chromatographischen Auftrennung von ^{14}C- und ^{32}P-markierten Stoffwechselprodukten. J Chromatogr 41:775–782

Fengler S, Spirer I, Neef M, Ecke M, Hauslage J, Hampp R (2015a) Changes in gene expression of *Arabidopsis thaliana* cell cultures upon exposure to real and simulated partial-g forces. Microgravity Sci Technol 28:319–329. https://doi.org/10.1007/s12217-015-9452-y

Fengler S, Spirer I, Neef M, Ecke M, Nieselt K, Hampp R (2015b) A whole-genome microarray study of *Arabidopsis thaliana* semisolid callus cultures exposed to microgravity and nonmicrogravity related spaceflight conditions for 5 days on Board of Shenzhou 8. BioMed Res Int 2015:547495. https://doi.org/10.1155/2015:547495

Forsthoefel NR, Cushman MAF, Cushman JC (1995) Posttranscriptional and posttranslational control of enolase expression in the facultative Crassulacean acid metabolism plant *Mesembryanthemum crystallinum* L. Plant Physiol 108(3):1185–1195

Gehring CA, Irving HR, Parish RW (1990) Effects of auxin and abscisic acid on cytosolic calcium and pH in plant cells. Proc Natl Acad Sci 87(24):9645–9649

Gilroy S, Bethke PC, Jones RL (1993) Calcium homeostasis in plants. J Cell Sci 106(2):453–461

Grant CM (2008) Metabolic reconfiguration is a regulated response to oxidative stress. J Biol 7(1):1

Gress TM, Hoheisel JD, Lennon GG, Zehetner G, Lehrach H (1992) Hybridization fingerprinting of high-density cDNA-library arrays with cDNA pools derived from whole tissues. Mamm Genome 3(11):609–619

Häder D-P, Hemmersbach R (1997) Graviperception and graviorientation in flagellates. Planta 203 (1):S7–S10

Halliwell B (2006) Reactive species and antioxidants. Redox biology is a fundamental theme of aerobic life. Plant Physiol 141(2):312–322

Hampp R, Goller M, Füllgraf H, Eberle I (1985) Pyridine and adenine nucleotide status, and pool sizes of a range of metabolites in chloroplasts, mitochondria and the cytosol/vacuole of Avena mesophyll protoplasts during dark/light transition: effect of pyridoxal phosphate. Plant Cell Physiol 26(1):99–108

Hampp R, Naton B, Hoffmann E, Mehrle W, Schönherr K, Hemmersbach-Krause R (1992) Hybrid formation and metabolism of plant cell protoplasts under microgravity. Physiologist 35(Suppl): S27–S30

Hampp R, Hoffmann E, Schönherr K, Johann P, De Filippis L (1997) Fusion and metabolism of plant cells as affected by microgravity. Planta 203(Suppl):S42–S53

Hampp R, Maier R, Martzivanou M, Ecke M, Magel E (2001) Gravitational effects on metabolism and gene expression of *Arabidopsis thaliana* cell cultures. In: Proceedings of 15th ESA symposium on European rocket and balloon programmes, Biarritz, pp 399–403

Hausmann N, Fengler S, Hennig A, Franz-Wachtel M, Hampp R, Neef M (2014) Cytosolic calcium, hydrogen peroxide and related gene expression and protein modulation in *Arabidopsis thaliana* cell cultures respond immediately to altered gravitation: parabolic flight data. Plant Biol (Stuttg) 16(Suppl 1):120–128

Hoffmann E, Schonherr K, Hampp R (1996) Regeneration of plant cell protoplasts under microgravity: investigation of protein patterns by SDS-PAGE and immunoblotting. Plant Cell Rep 15:914–919

Horneck G (1992) Radiobiological experiments in space: a review. Int J Radiat Appl Instrum D 20 (1):185–205

Horneck G (1999) Astrobiology studies of microbes in simulated interplanetary space. In: Laboratory astrophysics and space research. Springer, Dordrecht, pp 667–685

Hoson T, Soga K (2003) New aspects of gravity responses in plant cells. Int Rev Cytol 229:209–244

Hoson T, Soga K, Mori R, Saiki M, Nakamura Y, Wakabayashi K, Kamisaka S (2002) Stimulation of elongation growth and cell wall loosening in rice coleoptiles under microgravity conditions in space. Plant Cell Physiol 43(9):1067–1071

Hoson T, Soga K, Wakabayashi K, Kamisaka S, Tanimoto E (2003) Growth and cell wall changes in rice roots during spaceflight. In: Roots: the dynamic interface between plants and the earth. Springer, Dordrecht, pp 19–26

Johannes E, Collings DA, Rink JC, Allen NS (2001) Cytoplasmic pH dynamics in maize pulvinal cells induced by gravity vector changes. Plant Physiol 127(1):119–130

Karl T (2015) Microgravity-dependent changes of proteome phosphorylation in *Arabidopsis thaliana* cell cultures. Master thesis, University of Tuebingen

Kennedy AR (2014) Biological effects of space radiation and development of effective countermeasures. Life Sci Space Res 1:10–43

Kimbrough JM, Salinas-Mondragon R, Boss WF, Brown CS, Sederoff HW (2004) The fast and transient transcriptional network of gravity and mechanical stimulation in the Arabidopsis root apex. Plant Physiol 136(1):2790–2805

Kiss JZ (2000) Mechanisms of the early phases of plant gravitropism. Crit Rev Plant Sci 19 (6):551–573

Kiss JZ, Kumar P, Millar KD, Edelmann RE, Correll MJ (2009) Operations of a spaceflight experiment to investigate plant tropisms. Adv Space Res 44(8):879–886

Klychnikov O, Li K, Lill H, De Boer A (2006) The V-ATPase from etiolated barley (*Hordeum vulgare L.*) shoots is activated by blue light and interacts with 14-3-3 proteins. J Exp Bot 58 (5):1013–1023

Knight H, Trewavas AJ, Knight MR (1997) Recombinant aequorin methods for measurement of intracellular calcium in plants. Plant Mol Biol Manual C4:1–22

Kraft TF, van Loon JJ, Kiss JZ (2000) Plastid position in Arabidopsis columella cells is similar in microgravity and on a random-positioning machine. Planta 211(3):415–422

Krebs M, Held K, Binder A, Hashimoto K, Den Herder G, Parniske M, Kudla J, Schumacher K (2012) FRET-based genetically encoded sensors allow high-resolution live cell imaging of Ca^{2+} dynamics. Plant J 69(1):181–192

Lal SK, Lee C, Sachs MM (1998) Differential regulation of enolase during anaerobiosis in maize. Plant Physiol 118(4):1285–1293

Li J, Brader G, Palva ET (2004) The WRKY70 transcription factor: a node of convergence for jasmonate-mediated and salicylate-mediated signals in plant defense. Plant Cell 16(2):319–331

Limbach C, Hauslage J, Schafer C, Braun M (2005) How to activate a plant gravireceptor. Early mechanisms of gravity sensing studied in characean rhizoids during parabolic flights. Plant Physiol 139(2):1030–1040

Lin WH, Rui Y, Hui M, Xu ZH, Xue HW (2004) DNA chip-based expression profile analysis indicates involvement of the phosphatidylinositol signaling pathway in multiple plant responses to hormone and abiotic treatments. Cell Res 14(1):34–45

Link BM, Wagner ER, Cosgrove DJ (2001) The effect of a microgravity (space) environment on the expression of expansins from the peg and root tissues of Cucumis sativus. Physiol Plant 113 (2):292–300

Lu Y-T, Feldman LJ (1997) Light-regulated root gravitropism: a role for, and characterization of, a calcium/calmodulin-dependent protein kinase homolog. Planta 203:S91–S97

Maier R-M, Martzivanou M, Magel E, Zheng H, Cogoli-Greuter M, Hampp R (2003) Anaplerotic responses of "*Arabidopsis thaliana*" cell cultures to altered gravitational field strengths. In: European rocket and balloon programmes and related research, pp 253–258

Manzano AI, Van Loon JJ, Christianen PC, Gonzalez-Rubio JM, Medina FJ, Herranz R (2012) Gravitational and magnetic field variations synergize to cause subtle variations in the global transcriptional state of Arabidopsis in vitro callus cultures. BMC Genomics 13(1):105

Marco R, Leandro-Garcia LJ, Benguria A, Herranz R, Zeballos A, Gassert G, van Loon JJ, Medina FJ (2006) Gene expression variations during Drosophila metamorphosis in real and simulated gravity. In: 36th COSPAR scientific assembly, Beijing, 16–23 July

Martzivanou M (2004) Expressionsanalysen an Arabidopsis thaliana Kalluskulturzellen unter verschiedenen Gravitationsbedingungen. PhD thesis, University of Tübingen

Martzivanou M, Hampp R (2003) Hyper-gravity effects on the Arabidopsis transcriptome. Physiol Plant 118(2):221–231

Martzivanou M, Babbick M, Cogoli-Greuter M, Hampp R (2006) Microgravity-related changes in gene expression after short-term exposure of *Arabidopsis thaliana* cell cultures. Protoplasma 229:155–162

Miller G, Shulaev V, Mittler R (2008) Reactive oxygen signaling and abiotic stress. Physiol Plant 133(3):481–489

Minic Z, Jouanin L (2006) Plant glycoside hydrolases involved in cell wall polysaccharide degradation. Plant Physiol Biochem 44(7):435–449

Mittler R, Vanderauwera S, Gollery M, Van Breusegem F (2004) Reactive oxygen gene network of plants. Trends Plant Sci 9(10):490–498

Miyawaki A, Griesbeck O, Heim R, Tsien RY (1999) Dynamic and quantitative Ca^{2+} measurements using improved cameleons. Proc Natl Acad Sci USA 96(5):2135–2140

Moseyko N, Zhu T, Chang HS, Wang X, Feldman LJ (2002) Transcription profiling of the early gravitropic response in Arabidopsis using high-density oligonucleotide probe microarrays. Plant Physiol 130(2):720–728

Nagai T, Yamada S, Tominaga T, Ichikawa M, Miyawaki A (2004) Expanded dynamic range of fluorescent indicators for Ca^{2+} by circularly permuted yellow fluorescent proteins. Proc Natl Acad Sci USA 101(29):10554–10559

Nasir A, Strauch S, Becker I, Sperling A, Schuster M, Richter P, Weißkopf M, Ntefidou M, Daiker V, An Y (2014) The influence of microgravity on Euglena gracilis as studied on Shenzhou 8. Plant Biol 16(s1):113–119

Neef M, Hennig A, Hausmann N, Hampp R (2011) A parabolic flight profile as reflected by fluctuations in cytosolic calcium and gene expression of plant cells. In: 20th symposium on European rocket and balloon programmes and related research, pp 433–438

Neef M, Fengler S, Hausmann N, Ecke M, Hennig A, Franz-Wachtel M, Hampp R (2013a) Cytosolic calcium, hydrogen peroxide, and related gene expression and protein modulation in *Arabidopsis thaliana* cell cultures respond immediately to altered gravitation: a holistic view.

In: 21st ESA symposium on European rocket and balloon programmes and related research (ESA SP-721)
Neef M, Fengler S, Ecke M, Hausmann N, Hampp R (2013b) Joint european partial-g parabolic flight campaign calcium analysis in *Arabidopsis thaliana* cell cultures. In: Proceedings of 'Life in space for life on Earth' (ESA SP-706)
Neef M, Ecke M, Hampp R (2015) Real-time recording of cytosolic calcium levels in *Arabidopsis thaliana* cell cultures during parabolic flights. Microgravity Sci Technol 27(4):305–312
Neef M, Denn T, Ecke M, Hampp R (2016) Intracellular calcium decreases upon hyper gravity-treatment of *Arabidopsis thaliana* cell cultures. Microgravity Sci Technol 28:331–336
Neill SJ, Desikan R, Clarke A, Hurst RD, Hancock JT (2002) Hydrogen peroxide and nitric oxide as signalling molecules in plants. J Exp Bot 53(372):1237–1247
Obenland DM, Brown CS (1994) The influence of altered gravity on carbohydrate metabolism in excised wheat leaves. J Plant Physiol 144(6):696–699
Ong SE, Mann M (2006) A practical recipe for stable isotope labeling by amino acids in cell culture (SILAC). Nat Protoc 1(6):2650–2660
Ong SE, Blagoev B, Kratchmarova I, Kristensen DB, Steen H, Pandey A, Mann M (2002) Stable isotope labeling by amino acids in cell culture, SILAC, as a simple and accurate approach to expression proteomics. Mol Cell Proteomics 1(5):376–386
Orozco-Cárdenas ML, Narváez-Vásquez J, Ryan CA (2001) Hydrogen peroxide acts as a second messenger for the induction of defense genes in tomato plants in response to wounding, systemin, and methyl jasmonate. Plant Cell 13(1):179–191
Paul A-L, Popp M, Gurley W, Guy C, Norwood K, Ferl R (2005) Arabidopsis gene expression patterns are altered during spaceflight. Adv Space Res 36(7):1175–1181
Paul AL, Manak MS, Mayfield JD, Reyes MF, Gurley WB, Ferl RJ (2011) Parabolic flight induces changes in gene expression patterns in *Arabidopsis thaliana*. Astrobiology 11(8):743–758
Paul AL, Zupanska AK, Ostrow DT, Zhang Y, Sun Y, Li JL, Shanker S, Farmerie WG, Amalfitano CE, Ferl RJ (2012) Spaceflight transcriptomes: unique responses to a novel environment. Astrobiology 12(1):40–56
Perbal G, Driss-Ecole D (2003) Mechanotransduction in gravisensing cells. Trends Plant Sci 8 (10):498–504
Perera IY, Heilmann I, Chang SC, Boss WF, Kaufman PB (2001) A role for inositol 1,4,5-trisphosphate in gravitropic signaling and the retention of cold-perceived gravistimulation of oat shoot pulvini. Plant Physiol 125(3):1499–1507
Perera IY, Hung CY, Brady S, Muday GK, Boss WF (2006) A universal role for inositol 1,4,5-trisphosphate-mediated signaling in plant gravitropism. Plant Physiol 140(2):746–760
Perrin RM, Young L-S, Narayana Murthy U, Harrison BR, Wang Y, Will JL, Masson PH (2005) Gravity signal transduction in primary roots. Ann Bot 96(5):737–743
Plieth C, Trewavas AJ (2002) Reorientation of seedlings in the Earth's gravitational field induces cytosolic calcium transients. Plant Physiol 129:786–796
Poovaiah BW, McFadden JJ, Reddy AS (1987) The role of calcium ions in gravity signal perception and transduction. Physiol Plant 71:401–407
Poovaiah BW, Yang T, van Loon JJ (2002) Calcium/calmodulin-mediated gravitropic response in plants. J Gravit Physiol 9(1):P211–P214
Pyle T, Souret F, Correll M, Kiss J (2001) The effects of gravity and the spaceflight environment on the gene expression profile of *Arabidopsis thaliana* seedlings. Ph.D. thesis, Agricultural and Biological Engineering, University of Florida, Gainesville, FL, USA
Riechmann JL, Ratcliffe OJ (2000) A genomic perspective on plant transcription factors. Curr Opin Plant Biol 3(5):423–434
Robatzek S, Somssich IE (2001) A new member of the Arabidopsis WRKY transcription factor family, AtWRKY6, is associated with both senescence- and defence-related processes. Plant J 28(2):123–133
Roos W, Evers S, Hieke M, Tschope M, Schumann B (1998) Shifts of intracellular pH distribution as a part of the signal mechanism leading to the elicitation of benzophenanthridine alkaloids.

Phytoalexin biosynthesis in cultured cells of eschscholtzia californica. Plant Physiol 118 (2):349–364
Rosen E, Chen R, Masson PH (1999) Root gravitropism: a complex response to a simple stimulus? Trends Plant Sci 4(10):407–412
Sack FD (1991) Plant gravity sensing. Int Rev Cytol 127:193–252
Salmi M, Roux S (2008) Gene expression changes induced by space flight in single-cells of the fern Ceratopteris richardii. Planta 229(1):151–159
Sarwat M, Ahmad P, Nabi G, Hu X (2013) Ca(2+) signals: the versatile decoders of environmental cues. Crit Rev Biotechnol 33(1):97–109
Schütz W, Hausmann N, Krug K, Hampp R, Macek B (2011) Extending SILAC to proteomics of plant cell lines. Plant Cell 23(5):1701–1705
Shen-Miller J, Hinchman R, Gordon S (1968) Thresholds for georesponse to acceleration in gravity-compensated Avena seedlings. Plant Physiol 43(3):338–344
Sievers A (1991) Gravity sensing mechanisms in plant cells. Gravit Space Biol 4(2):43–50
Sobick V, Sievers A (1978) Responses of roots to simulated weightlessness on the fast-rotating clinostat. Life Sci Space Res 17:285–290
Soga K (2013) Resistance of plants to gravitational force. J Plant Res 126(5):589–596
Soga K, Kotake T, Wakabayashi K, Kamisaka S, Hoson T (2008) Transient increase in the transcript levels of gamma-tubulin complex genes during reorientation of cortical microtubules by gravity in adzuki bean (Vigna angularis) epicotyls. J Plant Res 121(5):493–498
Staves MP (1997) Cytoplasmic streaming and gravity sensing in Chara internodal cells. Planta 203 (Suppl 1):S79–S84
Steingraber M, Outlaw WH, Hampp R (1988) Subcellular compartmentation of fructose 2, - 6-bisphosphate in oat mesophyll cells. Planta 175(2):204–208
Stracke R, Werber M, Weisshaar B (2001) Th.E R2R3-MYB gene family in *Arabidopsis thaliana.* Curr Opin Plant Biol 4(5):447–456
Stutte G, Monje O, Hatfield R, Paul A, Ferl R, Simone C (2006) Microgravity effects on leaf morphology, cell structure, carbon metabolism and mRNA expression of dwarf wheat. Planta 224(5):1038–1049
Sun L, van Nocker S (2010) Analysis of promoter activity of members of the PECTATE LYASE-LIKE (PLL) gene family in cell separation in Arabidopsis. BMC Plant Biol 10(1):152
Takeda S, Gapper C, Kaya H, Bell E, Kuchitsu K, Dolan L (2008) Local positive feedback regulation determines cell shape in root hair cells. Science 319(5867):1241–1244
Toyota M, Gilroy S (2013) Gravitropism and mechanical signaling in plants. Am J Bot 100 (1):111–125
Toyota M, Furuichi T, Tatsumi H, Sokabe M (2008) Cytoplasmic calcium increases in response to changes in the gravity vector in hypocotyls and petioles of Arabidopsis seedlings. Plant Physiol 146(2):505–514
Toyota M, Ikeda N, Sawai-Toyota S, Kato T, Gilroy S, Tasaka M, Morita MT (2013) Amyloplast displacement is necessary for gravisensing in Arabidopsis shoots as revealed by a centrifuge microscope. Plant J 76:648–660. https://doi.org/10.1111/tpj.12324
Tuteja N, Mahajan S (2007) Calcium signaling network in plants: an overview. Plant Signal Behav 2(2):79–85
Umeda M, Manabe Y, Uchimiya H (1997) Phosphorylation of the C2 subunit of the proteasome in rice (*Oryza sativa L.*). FEBS Lett 403(3):313–317
Vasilenko A, Popova A (1996) Energetic metabolism response in algae and higher plant species from simulation experiments with the clinostat. Adv Space Res 17(6–7):103–106
Walther I, Pippia P, Meloni MA, Turrini F, Mannu F, Cogoli A (1998) Simulated microgravity inhibits the genetic expression of interleukin-2 and its receptor in mitogen-activated T lymphocytes. FEBS Lett 436(1):115–118
Wan XS, Zhou Z, Kennedy AR (2003) Adaptation of the dichlorofluorescein assay for detection of radiation-induced oxidative stress in cultured cells. Radiat Res 160(6):622–630

Wan XS, Zhou Z, Ware JH, Kennedy AR (2005) Standardization of a fluorometric assay for measuring oxidative stress in irradiated cells. Radiat Res 163(2):232–240

Wang H, Zheng HQ, Sha W, Zeng R, Xia QC (2006) A proteomic approach to analysing responses of *Arabidopsis thaliana* callus cells to clinostat rotation. J Exp Bot 57(4):827–835

Wong HL, Pinontoan R, Hayashi K, Tabata R, Yaeno T, Hasegawa K, Kojima C, Yoshioka H, Iba K, Kawasaki T (2007) Regulation of rice NADPH oxidase by binding of Rac GTPase to its N-terminal extension. Plant Cell 19(12):4022–4034

Yoshioka R, Soga K, Wakabayashi K, Takeba G, Hoson T (2003) Hypergravity-induced changes in gene expression in Arabidopsis hypocotyls. Adv Space Res 31(10):2187–2193

Zhang Y, Wang L, Xie J, Zheng H (2015) Differential protein expression profiling of *Arabidopsis thaliana* callus under microgravity on board the Chinese SZ-8 spacecraft. Planta 241 (2):475–488

Zupanska AK, Denison FC, Ferl RJ, Paul A-L (2013) Spaceflight engages heat shock protein and other molecular chaperone genes in tissue culture cells of *Arabidopsis thaliana*. Am J Bot 100 (1):235–248

The manufacturer's authorised representative in the EU is Springer Nature Customer Service Centre GmbH, Europaplatz 3, 69115 Heidelberg, Germany. If you have any concerns regarding our products, please contact ProductSafety@springernature.com

Printed and bound by CPI Group (UK) Ltd, Croydon, CR0 4YY
01/10/2026
02225880-0009